A History of the "New Mathematics" Movement and its Relationship with Current Mathematical Reform

Angela Lynn Evans Walmsley

T0127892

University Press of America,® Inc.
Lanham · New York · Oxford

Copyright © 2003 by
University Press of America,® Inc.
4501 Forbes Boulevard, Suite 200
Lanham, Maryland 20706

PO Box 317
Oxford
OX2 9RU, UK

ISBN 0-7618-2511-8 (clothbound : alk. ppr.)
ISBN 0-7618-2512-6 (paperback : alk. ppr.)

For Stephen and Patrick

Contents

Preface

This book was created because it did not exist – a history of the "new mathematics" movement has never been published in its entirety. As a graduate student I became intrigued with the history of mathematics education, but became particularly interested in history that seemed to be repeating itself. The "new math" movement is an important and historical mathematical movement in our history that was suddenly abandoned. However, understanding and learning from the past is crucial for mathematics teachers and educators; and referring to what we have done in the past as it relates to today is beneficial to teaching.

In this book, I attempt to provide a thorough history of the "new mathematics" movement as it occurred in the United States. I provide an historical framework first so that the reader can understand the "new mathematics" movement in context with history. Then I explain the "new mathematics" movement based on the many projects that were developed at the time. The combination of these projects is what defines the content and actions of the "new math" movement. By doing this, it is my hope that educators can develop a true understanding of what really happened during the "new math" era. I also offer suggestions as to why the "new math" movement was seen as failed.

I have tied the "new math" movement to the current mathematical reform movement led by the National Council of Teachers of Mathematics (NCTM). Because the "good things" from the "new math" movement are so similar to what we are doing today, the reader can see the relationship in this book and use the information wisely when developing and improving his or her own teaching. Some of the things we as educators have learned from the "new math" movement can help us help our students without the same mistakes made then.

It is my hope that this book will help us all move forward during these difficult times in education to serve our students better. Thank you to the many teachers, administrators, educators, and students of the "new math" movement. Because of your vision and experimentation, we can educate our current and future students better as we strive for perfection in mathematics teaching and learning.

Introduction

The "new mathematics" movement can be defined as a generic term referring to the mathematical reform effort in the United States during the 1950s and 1960s whose leaders desired reforming, enhancing, or repairing mathematics education. These reform efforts were most often established by university or government mathematical projects designed to develop "new mathematics" curricula. The main goal of the authors of the "new math" programs was to present mathematics as a logical structure to children who could then develop an understanding and appreciation for mathematical reasoning and mathematical principles[1]. The "new mathematics" movement is often referred to as the "new math" or "modern mathematics." While the reform movement did offer some "new" content, the reader must be aware that "new mathematics" was a mixture of both new and old ideas[2].

The "new mathematics" movement was a powerful movement in the history of mathematics education because it caused the American public to focus on improving mathematical knowledge for all students. The movement brought to schools across America a new curriculum, new pedagogical techniques, and higher standards for teachers and students. While many felt this new approach to mathematics faded away with the early 1970s "Back to Basics" movement due to its failure to improve overall mathematical ability in students, little evidence supports this. In fact many elements of the "new mathematics" reform are present in the current mathematical reform movement led by the National Council of Teachers of Mathematics with their first publication of the *Curriculum and Evaluation Standards for School Mathematics* in 1989 and second publication titled *Principles and Standards for School Mathematics (PSSM)* in 2000.

Very little literature has been written concerning the "new mathematics" movement with regards to the current mathematics curriculum. In fact, most literature pertaining to the "new mathematics" movement was written during the movement or shortly afterwards—without the knowledge of post "new mathematics" reform. To understand current reform, the National Council of Teachers of Mathematics (NCTM) must be consulted. NCTM is the major professional organization for school mathematics teachers in the United States and a leader in modern mathematics reform. When the author contacted NCTM about this topic, the only publication suggested was the 32nd NCTM Yearbook, *A History of Mathematics Education in the United States and Canada*, published in 1970. While this book is helpful, it was published at the end of the "new math" movement without the opportunity to reflect on the movement with an overall perspective. More surprising, however, is that there exists no comprehensive analysis of the "new math" movement in a publication since. Therefore, the author was encouraged to explore the "new math" era because of the need for and the lack of research in this important reform.

The history of the "new math" movement—particularly in light of present reform—is an important topic not yet researched. While some articles and books speak of the "new mathematics," few provide a comprehensive understanding of the history of "new mathematics," and none provide an unbiased view. Another issue with information to date concerning "new math" is that most provides some kind of bias because it has been completed either by individuals directly involved in the production or teaching of "new math" curricular materials or individuals who were students at the time of the "new math" movement. While their views may be positively or negatively biased, the fact that they were present during the movement means that they were affected in some way differently than someone who can approach the "new math" movement objectively. The author has provided a less biased view of connections between "new math" and current mathematical reform.

The relationship between "new math" and present reform is an important topic for mathematics educators. This book offers a contribution to the mathematical community by establishing a thorough knowledge of the history of the "new mathematics" movement and substantiating its influence on modern educational reform. Lack of connections between the "new math" movement and the present mathematical reform has been a disservice to mathematics educators because many risk repeating similar efforts without the knowledge of the past reform efforts or these connections. Before discussion of the "new math" can take place, this book provides an historical understanding of the "new math" and its relationship to current reform so that mathematics educators can learn from history rather than repeating it unnecessarily.

Much of the general public believes that the history of the "new math" began with the launch of the first artificial satellite, *Sputnik*, on October 4, 1957, by the USSR[3]. However, mathematics reform was actually debated in the 1930s and 1940s, and "new math" began in earnest in the early 1950s with the start of many mathematics projects led by university mathematicians throughout the United States. The National Science Foundation, established in 1950 by an act of Congress, contributed greatly to the mathematics projects of the 1950s and 1960s. The sudden amount of federal funds that became available through the United States National Defense Education Act of 1958, partly as a result of *Sputnik*, made reform an inevitable reality. By the mid 1960s, the reform of secondary mathematics education for college bound students was well established, and many educators began to look at how children learn mathematics and how this affected pedagogy.

The launching of *Sputnik* shook the nation's faith in the American school system, especially in the areas of science and mathematics. Americans saw their national safety threatened because the USSR appeared to be more advanced technologically than the United States. "Sputnikshock" brought the problems of a low standard of mathematics education to the public. Many agreed that previous teaching of mathematics had been unsuccessful, and reform needed to take place[4]. Developing a modern mathematics curriculum was the reform tool chosen, as though it could solve the nation's mathematical ills. "New math" became widespread because of the large number of forces that happened to push for it at the same time[5]. However, while most people know *of* the "new mathematics," many do not know *what* comprised the "new mathematics."

The "new mathematics" reform offered new content as well as new teaching approaches or pedagogy[6]. The new content normally consisted of abstract algebra, topology, symbolic logic, set theory, and Boolean algebra, and was taught in conjunction with much of the traditional curriculum[7]. Set theory and general principles of modern algebra made the "new mathematics" more abstract than the traditional mathematics[8]. Set theory often became synonymous with "new mathematics." The general public thought set theory (the idea of a collection of similar items) was being taught instead of arithmetic[9]. The conventional curriculum relied on drill and practice with little emphasis on understanding. The main idea of "new mathematics" was to abandon rote learning and teach with a more logical approach to mathematics where students could *understand* mathematics. The reform to develop a new school curriculum was led by college mathematicians[10]. This reform took place in the form of mathematics projects and committees that developed new curricular materials. These materials were widely distributed and used in various schools throughout the United States; unfortunately many of the materials were used by teachers with little training in "new mathematics" concepts. As a result, children were learning more abstract concepts without an apparent knowledge of basic skills. Teachers and parents became frustrated with understanding and explaining the content of the "new mathematics" curricula. Overall, the general public began to lose interest in the "modern mathematics" movement as it was not providing the results they had hoped.

Therefore, the "new mathematics" movement began to disappear in the early 1970s. The beginning of this decade brought a backlash to the "new mathematics" and a return to emphasis on paper and pencil skills. This era has often been named "Back to Basics"[11]. Near the end of the 1970s, mathematics reform turned towards an emphasis on problem solving. Most educators were content with the advancement of basic skills, but they wanted to concentrate on teaching students how to solve problems in an applied context[12]. The National Council of Teachers of Mathematics published a document during the 1980s entitled *Agenda for Action* in which they stressed the need for problem solving. At the beginning of the 1990s, mathematics reform began to include other topics such as technology and mathematical modeling which represented problem solving skills in "real life" situations.

The largest driving force for mathematics reform in the 1990s was the National Council of Teachers of Mathematics and its publication of *Curriculum and Evaluation Standards for School Mathematics*. This book, often referred to simply as the *Standards,* makes recommendations about what mathematics students should learn in grades kindergarten through twelve. The *Standards*, published in 1989, stressed the importance of problem solving and reasoning as well as basic skills. However, NCTM also encourages the use of higher levels of mathematics when appropriate. The second edition of the *Standards* was published in 2000. *Principals and Standards for School Mathematics* (often referred to as the *Standards* or *PSSM)* shows the acceptance of the first edition and the importance of the *Standards* to many mathematics educators and NCTM. The *Standards* discuss teaching and learning strategies that enable students to work at their best abilities in mathematics[13]. The *Standards* has become a popular reference guide for many mathematics educators in the United States. (An example of the content of the *Standards and PSSM* can be found in Appendices A and B). Because of the widespread knowledge of these guides by NCTM, they should be used when making connections with the "new math" influence.

The "new mathematics" era was a time filled with energy, enthusiasm, and idealism. Many felt that "modern mathematics" was an improvement on the old curriculum, but the value and success of the "new mathematics" era fell short of what was expected. Parents and educators did not see as large an increase in mathematical knowledge as they had expected. However, whether this "increase in knowledge" was ever accurately measured or tested is a valid issue pertaining to the "new math" movement. "New mathematics" is known to most as a failed movement, and this opinion has hindered the country's ability to learn from history and past mathematics reform. "New mathematics" ended when those who brought it into being ceased producing materials and supporting the movement. Conflicts arose between the pioneers in curriculum reform and other groups such as the general public and government who were interested in mathematics reform. While rapid and widespread reform of mathematics teaching was a goal of the public and government, learning modern mathematics was not. The era of "new mathematics" ended also in part because of the changing priorities of our country:

> The public, which had seen technology and keeping ahead of the Russians as major goals, was beginning to question our role in Vietnam and the impact of technology on the environment. The U.S. had caught up in the space race just as public interest in that race began to decline. Young people turned from academic or career interests towards personal growth, spiritual

awareness, and drugs. The disappearance of all the vital forces that brought the 'new math' into being left it an empty shell that crumbled rapidly under the attack of disgruntled parents and teachers. In addition, the disappearance of those forces left the 'new math' relatively unsupported and undefended, making its 'failure' appear, perhaps, much greater than it was[14, p.7].

The powerful and influential "new mathematics" movement has affected mathematics education; however, no one is quite sure as to what the effects have been. Therefore, the connections between the "new mathematics" movement and the *Standards* are explored in this book. In order to explore these connections, a substantial historical analysis of the "new math" movement is necessary. Included are topics or issues such as the influence on pedagogy, the influence on the curriculum, the influence on goals, the influence on assessment, and the influence on educational standards—all topics addressed within the *Standards*. By analyzing the history of the "new mathematics" movement within the context of the history of the United States, the author has provided information pertaining to the factors that have framed the *Standards*. The history of the "new mathematics" movement began in the early 1950s and ended in the late 1960s or early 1970s. Therefore, the history of the movement is analyzed within this time frame. To understand the relationship to the *Standards*, the *Standards* is also analyzed, providing a background for reform between it and the "new math" movement. The author has discussed the influence of the "new mathematics" movement not only in the context of the history of the United States, but also in the context of the history of mathematical and educational reform.

Chapter Two addresses the history of the "new mathematics" movement by describing the general societal view of education at the time. Many important issues in history are addressed in order to provide a framework for the reader to understand why some of the mathematical changes were taking place. Some of these issues include the lack of mathematical ability evident in World War II, the launch of *Sputnik,* the assassination of President Kennedy, and the "space race." The author has also discussed the teacher and student standards within the context of the mathematical curriculum as well as the pedagogical techniques evident before and during the "new mathematics" movement.

Chapter Three provides a comprehensive analysis of funded mathematics projects and influential publications during the "new mathematics" era. The major projects and committees who developed a new mathematical curriculum are discussed in chronological order. Next, smaller projects are described. Similarities between these projects are discussed after the explanations of the different projects.

Chapter Four focuses on the educational changes in the learning of mathematics of students and teachers during the "new math." Some explanations of the disappearance of the "new math" are also discussed. A short explanation of reform from the mid 1960s to the publishing of the *Standards* in 1989 is provided.

Chapter Five discusses the connections of the "new mathematics" movement with the *Standards*. The goals and content of the *Standards* as well as the similarities and differences between the "new math" movement and the *Standards* movement are discussed in five different themes: pedagogy, philosophy, content, teaching, and assessment. Included in this chapter is an explanation of pedagogical techniques and teacher and student standards presently expected in light of the connections from the "new math" era.

Chapter Six summarizes the "new mathematics" movement and its effects on present day mathematics education reform. The author has also provided recommendations for future mathematics reform concerning curriculum and pedagogy in light of both the "new mathematics" movement and the *Standards* movement.

"New Mathematics" History

Chapter Two begins with a history of the many events that took place in the United States surrounding the "new mathematics" movement. In order to understand mathematical reform, the reader must have a context for why changes in mathematics were being made. Therefore, various historical and educational issues of the early 1950s through the late 1960s must be addressed—beginning with the effect of World War II preceding the "new math" movement. Other important issues that will be discussed in this chapter are the launching of the Russian satellite, *Sputnik,* and the entire "space race" and space program. Other factors include the election and later assassination of President Kennedy. Also, an important issue that will be addressed is the general attitude towards education by the American society at the time. An historical framework for the reader to use in understanding some of the changes in mathematical reform during the "new mathematics" movement is provided.

Equally important for the reader to understand were the current mathematical curricula and practices established prior to the "new mathematics" movement as well as how the curriculum and practices changed during the movement. Therefore, the traditional mathematical curriculum (including both teacher and student standards) and pedagogical techniques will be discussed. While some of the mathematics projects that will be discussed in the next chapter touch on some of these issues, a general explanation of the mathematics curriculum and changes in pedagogy is crucial in order to show the relationships of these to the *Standards*.

Historical Context

Before and during World War II, many German mathematicians immigrated to the United States as Hitler rose to power[1]. Their own background was solid in pure and applied mathematics because of their rigorous education in Europe. In general, they had a broader view of mathematics related to both philosophy and science. When called upon to train military soldiers, the evidence of weakness in mathematics of the recruits was apparent to these mathematicians[2]. The United States was beginning to see the need for strong mathematics in advancing technology and maintaining military strength throughout the war—particularly with the advancement of the programmable computer and operations research[3]. Operations research is often credited with winning World War II. Operations research is loosely defined, but it describes the use of science and mathematics in deciding the procedures in war using available weapons and manpower[4]. Electronic computers were being developed along the same lines—to help in utilizing weaponry to the fullest extent. These mathematicians made the public and government aware of the contributions science and mathematics could make to the military strength and economic development of a country.

While the military and government were becoming very aware of the importance of mathematics for military strength, the nation as a whole was unaware. As World War II developed, the general public soon understood that in order to develop military technology and industrial production, we must place emphasis on mathematics and science. The importance of science and mathematics was very clear in the development of the atomic bomb—the product seen as causing the end of the war. Consequently, science and mathematics research was given continued funding by the government.

Furthermore, a number of mathematicians returned to their colleges and universities at the end of World War II to resume their teaching duties. During the war they had been exposed to developments in mathematics such as linear programming, game theory, and new methods in statistics[5]. Many found the old mathematics dull and not connected to the modern mathematics apparent during the war. Therefore, they desired a change in mathematics taught at the university level to include more modern concepts.

On September 6, 1945, President Truman sent the "Fair Deal" to Congress. With the end of World War II, the President was hoping to focus on some of the problems in the United States. Among other issues, President Truman requested money for education and scientific research; however, Congress did not act on this issue[6]. After the end of World War II, the United States could no longer slip back into its own world—it would forever be involved in conflict in Europe[7]. Entrance into the North Atlantic Treaty Organization (NATO) meant the United States needed to have an organized military in peacetime; therefore involving itself in European politics and economies. The United States had become part of a world patrol force. While accepting global burdens, the US was committed to strengthening military intelligence[8]. Therefore, Congress felt money needed to be placed in foreign affairs and military strength instead of education.

However, education slowly began to obtain more attention by the nation. Sources such as the National Science Foundation were providing mathematics and science research throughout universities in the United States after the creation of the National Science Foundation in 1950. With the return of many soldiers to the US after the end of WWII, colleges and universities were filled with those who took advantage of the GI Bill of Rights[9]. There was general agreement in the early 1950s that Americans performed poorly in mathematics, which was especially evident in World War II. Thus, the need to improve mathematics among all citizens became imperative. The Department of Defense stated in 1957 that they were advancing faster than researchers or universities with the development of scientific and technological machines such as computers, electronic controls and automation as they saw further technological development essential after WWII[10].

While Eisenhower was concerned with social welfare and internationalism, the Cold War was in full swing by 1950 causing much concern and focus once again on foreign affairs rather than problems within education. While the United States worked with the Soviet Union in some international policies in the 1950s, the United States was still wary of the USSR. Some mathematics educators were afraid that the United States was not encouraging mathematics and science enough to counteract the forces of Russia[11]. Yet the public, while they supported mathematics, did not feel in danger of technology superiority by any other nation. In August of 1957, the USSR announced that it had tested an intercontinental ballistic missile—a feat the US had not achieved yet[12]. While the general public of the United States was not completely in fear of "lagging behind" the USSR, this view completely changed with the launching of *Sputnik*.

When the USSR launched the first satellite, *Sputnik*, on October 4, 1957, technological advances appeared to be tipped towards the USSR. The Soviet Union launched a second *Sputnik* in October as well showing increased success and achievement[13]. Many believe this date marked the beginning of the "space race" between the US and the USSR. The general public was displeased with President Eisenhower, a military man, for allowing the Russians to "get ahead" in the "space race." However, President Eisenhower said that Russian and American space programs were never considered a race[14]. Wooten describes the effect of *Sputnik*:

> Additional impetus was given to the reform efforts when on October 4, 1957, the Soviet Union launched Sputnik I and injected the factors of national prestige and national security into the picture. This technological achievement raised questions regarding the mathematics curriculum in the United States that carried the controversy out of the world of scholars and into the public domain[15, p.9].

The launching of *Sputnik* did not actually mean that the USSR was "ahead." However, it definitely gave that appearance to the general public. In retrospect, the United States had superior military technology and nuclear weaponry, and the US launched its first satellite on January 31, 1958[16]. However, none of these issues appeased the general public, and a push in schools for more academics began as the education system became the scapegoat for the problem. Therefore, the launching of *Sputnik* became the most influential promoter of mathematics education in the United States. At this same time, through popular magazines, the public was just beginning to hear about the "new math" movement that was taking shape. Local newspaper articles, national newspaper articles found in newspapers such as *The Washington Post and the New York* Times, and popular magazine articles found in magazines such as *Reader's Digest, Life, and Newsweek* showed that the "new math," while different for many parents and readers, was favorably accepted at the beginning of the movement[17].

The National Science Foundation (NSF) had been established in 1950 with $15,000,000; however, after *Sputnik*, their annual fund was raised to $140,000,000. Waterman, the director of the NSF, described the *Sputnik* crisis as a "scientific Pearl Harbor" because of the public panic and outcry and the immediate reactions by the government[18, p.224].

The National Science Foundation formed as a means to support university scientific research as well as promote scientific research nationally. Because many universities were losing valued researchers to industry, the NSF hoped to provide support to strengthen the education of future scientists[19]. The NSF provided monies for fellowships for advanced study in mathematics and science; improvement of K-12 mathematics, science, and foreign language; more vocational programs; and guidance counseling as well as testing of gifted students[20]. Much of the NSF funding was established for multiple summer teaching institutes specifically for secondary mathematics and science teachers. When the NSF attacked the problem of educating the many elementary teachers, they realized that it would be impossible to reach a majority because in effect, all elementary teachers were science and mathematics teachers. They did offer many institutes, but never achieved continued education for a large percentage of the elementary school teaching population[21].

The National Defense Education Act (NDEA) was passed in 1958 which gave approximately one billion dollars to be spent over four years to promote mathematics, science, and foreign language[22]. This act did not address quality of education, but instead was an anxious move by Congress following *Sputnik* to improve college-level education—particularly in applied science and engineering. The NDEA gave money to colleges in the form of grants, but a majority went directly to students in the form of loans, scholarships, and fellowships. The College Entrance Examination Board published a report in 1959 that warned that mathematics curricular reform must be valued for the sake of improving mathematical ability rather than as a by-product of the post-Sputnik panic[23].

During 1958, the National Aeronautics and Space Administration (NASA) was established[24]. Despite the actual ability of the USSR and the technology which has been considered equal to or less than the technological abilities of the US at the time, the "space race" had begun with full force. In fact, when President Kennedy was elected in 1960, he made the "missile gap" a major part of his campaign, announcing that the U.S. would send a man to the moon before the end of the decade. Public interest in mathematics was at an all time high in 1958, and curriculum reform was in process or in planning stages all over the country[25].

With Kennedy's election in 1960, America had a new "all-American" icon and a general hope for a better future. Kennedy was a young, vibrant President who wanted to improve the US economy and make a positive mark in foreign policy. He called for a "New Frontier." He started the Peace Corps and requested a more mobile and technically skilled armed forces[26]. However, the 1960s decade proved to be much more turbulent than expected on a number of issues.

The poor of America suffered during the 1950s as the economy did not fare well, and many citizens ignored the poor which included many minorities. The Eisenhower administration had allowed welfare to become a permanent state with little other focus on the poor. The end of the 1950s saw an interest in more social issues such as civil rights, safety, air and water pollution, and better schooling. With the Supreme Court being led by Justice Warren (1953-1969), civil rights were finally being addressed. The court presided over *Brown vs. Board of Education* in 1954 which denounced the separate but equal clause in *Plessy vs. Ferguson*. Furthermore, Eisenhower backed federal laws enforcing integration by signing the Civil Rights Act of 1957. However, major changes for the poor and minorities did not occur until Kennedy's administration.

The early 1960s incurred much unrest concerning civil rights, and Kennedy started attacking unfair laws in 1963. Kennedy also wanted new legislation that would increase the quality and availability of education for all. Kennedy asked for a National Education Improvement Act, but Congress never supported him in this idea. With the assassination of President Kennedy in November of 1963, some of the hopefulness for a better country died as well. However, President Lyndon Baines Johnson asked Congress to honor President Kennedy by supporting the changes he was trying to make in the "New Frontier".

Therefore, during LBJ's presidency, more laws were passed than in any earlier era. These laws focused on civil rights, health, education, eradication of poverty, and aid for cities. Johnson carried social plans far and started his own "War on Poverty." After his election in 1964, he showered Congress with Great Society legislation providing aid to education and healthcare[27]. The Civil Rights Act of 1964 was also passed which enforced the *Brown vs. Board of*

Education decision. In 1965, the Elementary and Secondary Education Act gave aid to poverty-stricken schools[28]. The five main titles passed under this law were issues such as grants given to local public school districts in aid of disadvantaged students; an increase in school library resources; money for new teaching methods and special projects; grants for the support of research in education; and money given for the development of leadership in state education departments.

However, as the 1960s progressed, political unrest continued—particularly on college and university campuses. College students were uninterested in a life of security and materialism. Instead, they were involved in creating a more fair and idealistic world. The emphasis was shifting away from academic excellence to educational opportunity equality[29]. Students were interested in helping the less fortunate, and the civil rights movement brought out the unfairness to minorities as well as the poverty in the United States. This unrest on college campuses led to the development of a counterculture which began during the civil rights demonstrations at the University of California at Berkeley in 1964. Problems on campuses continued as students protested the increased involvement of the United States in the war in Vietnam which was a major conflict by 1965. The pressures of the anti-Vietnam war movements also affected schools because school reform was not seen as "in line" with Johnson's Great Society program. Darkness fell on the mood of most Americans with the assassination of Martin Luther King, Jr., in April 1968 followed by the assassination of Robert Kennedy in June 1968 as well as the riots at the Democratic National Convention in Chicago also in 1968. Many members of this counterculture turned to drugs, sexual experimentation, and protests because of the increased violence within the country climaxing with the four fatal shootings at Kent State University in 1970 by the National Guard.

Johnson, who advocated his "war on poverty" near the beginning of the 1960s actually turned his focus away from the problems expressed by the counter culture towards political issues. Johnson began focusing more on the war in Vietnam by 1968 when he announced he would not run for re-election. Vietnam became Nixon's War after Johnson passed it to him in 1968. Besides the focus on the military and war, the "space race" ended when the United States was the first to place men on the moon in 1969. Some call this "cultural lag" because technology had advanced enough to send someone to the moon yet America seemed to have too many poverty and peace problems. While many agreed that the moon landing was a tremendous achievement, much of society had redirected its focus away from technological advances and had lost interest in the "space race." Instead, the general public wanted to focus on the racial and poverty problems within the United States.

General Curriculum

Concerning the education curriculum in general, James B. Conant wrote *The American High School Today*, based on his study of many various high schools in the United States. His book was published in 1959—giving a vivid picture of high schools and his recommendations as the "new math" movement was moving forward. Conant did not believe in comparing American schools with Russian schools as had been the case by previous researchers[30]. Instead, he focused on stating the strengths of American schools as well as offering a number of recommendations.

Conant explained that the US had some specialized schools such as vocational schools or college-preparatory schools, but that the vast majority were comprehensive schools—both large and small. His study showed that only a small number of boys and girls were given the opportunity to excel in an advanced program of study concerning mathematics, science, and foreign language. He referred to those who studied these three subjects as the "academically talented." Overall, Conant felt the academically talented youth of American culture were not being challenged enough. When asked about their top 15% – 20% students, most schools did not think they were enrolling a majority of them in twelfth grade mathematics (often trigonometry) or physics. Conant described the academically talented as the top 15% of the high school population nationally, and that they must be talented in both mathematics and foreign language. Students were often taking "easy classes" as opposed to challenging themselves in order to gain better grades and stay at a higher rank in their class.

Therefore, Conant stressed that America was ignoring its talented and gifted students by not sufficiently challenging them, not requiring them to work hard enough, and not offering subjects in a sufficient range. Furthermore, the balance between encouraging both academically talented boys and girls was unequal—with more boys pursuing advanced mathematics and science courses, and more girls pursuing foreign language. Conant stated this was a disservice to the students as well as the nation as it is in the national interest to develop the gifted to become the professional leaders of the future. He recommended that all academically talented students take four years of mathematics, four years of one continual foreign language, four years of English, three years of social studies, and a few elective courses. Conant also stated that the highly gifted students (approximately 3% of the population) should have special arrangements such as an advanced placement program in order to fully develop their academic potential. Also, summer school should be offered to both "slow" and "bright" children.

While Conant continued to stress the importance of challenging the academically talented, he also stressed educating all students to their full potential. He recommended all children have four years of English, three to four years of social studies, at least one year of algebra or general math in the ninth grade, and one year of biology or general science in the tenth grade. However, Conant explained that students should be ability grouped and that at least two different science and math courses should be offered at each level—therefore, he encouraged all students to pursue science despite whether or not they would be enrolled in the math-based science or the general science courses. NCTM, in *The Revolution in School Mathematics*, stated as well that students can learn much more mathematics earlier than ever was expected, but also agreed that students must be ability grouped to do so[31].

Conant specified three main objectives of comprehensive schools as: 1) to provide a solid general education for all students; 2) to provide skills that students could use in a job immediately following graduation; and 3) to provide programs for college-bound students that were appropriate for their future education in a university[32]. Furthermore, counselors should be made available to encourage and guide students in their mathematics choice. Price stated that all college-bound students should be required to take three years of "good" mathematics in school, but those who have mathematical interests and abilities should take four years of "good" mathematics[33]. One of the biggest problems tracking academically talented students was in the large number of small high schools in the United States with only one or two mathematics teachers. In 1952, 55% of the public high schools in the US had less than 200 students[34]. One teacher would find it almost impossible to educate two tracks of mathematics; thus, the number of small schools may have led to problems in identifying and adequately challenging talented students in those schools.

Conant stressed that the secondary schools of the United States must provide a solid mathematical background for all students while improving and training the academically talented. Conant was not alone in thinking that America was ignoring its gifted children. G.B. Price argued in 1951 that ignoring the gifted children was ignoring the need of talented people for national defense purposes[35]. As there was a shortage of school mathematics teachers, engineers, technicians, and research scientists, the US had to start recognizing its gifted students and encouraging mathematics for further study[36].

Mathematics Curriculum

Prior to World War II, most mathematics offered in secondary schools in the United States was based on mathematical procedures pertinent to business, government, and industry reflecting the society[37]. Most mathematics taught in elementary schools consisted of arithmetic skills taught by rote and drill[38]. Thus, most mathematics throughout the K-12 curriculum was traditionally taught with an emphasis on drill and practice. Many educators had the desire to bring the curriculum "up to date" as they complained that the conventional curriculum was based on mathematics topics developed hundreds of years ago[39]. The traditional mathematics curriculum in a medium to large high school was based on two tracks: one that was for the college bound yet did not challenge the students to their full potential, and another that was for the non-college bound yet did not offer them content at a higher level than arithmetic[40].

Shortly before World War II, the National Council of Teachers of Mathematics and the Mathematical Association of America worked together to produce a guidance report for the future of mathematics education in the United States[41]. This report stated that students in elementary grades should study the fundamentals of arithmetic including the four basic operations (addition, subtraction, multiplication, and division), fractions and decimals, measurement, estimation, and some shape. This describes the elementary education curriculum prior to the "new math" movement. Stress was placed on teaching arithmetic at a high standard.

The high school mathematics curriculum was generally offered in a "double track" system as mentioned previously. The college bound track was outlined by grade level with algebra or general mathematics taught in ninth grade, plane geometry taught in tenth grade, solid geometry taught in eleventh grade, and trigonometry taught in twelfth grade; or algebra taught for two years, plane geometry taught one year, and solid geometry combined with trigonometry taught another year. This curriculum was not the standard in all schools, but was the set curriculum in larger schools with a significant college-bound population of students. Many schools, particularly smaller high schools, did not offer a mathematics course in the last two years of school[42]. Many students preparing for college also did not take the last two years of mathematics classes offered in high schools because they had already fulfilled the state requirements for entrance to college[43]. Furthermore, some advanced mathematics was not actually much more advanced than previously offered courses. For example, while an intermediate algebra class claimed to be "intermediate," it often was almost a complete review of elementary algebra. While some schools may have also offered some analytic geometry, calculus, and statistics, the curriculum described above explains the typical high school curriculum before the "new math" movement.

The "other track" normally offered in large schools included courses like general mathematics followed by a consumer or "shop" mathematics or possibly some statistics[44]. This joint report published before World War II also stated that schools must provide a solid mathematics background for all students while pushing the academically talented further. Issues concerning learning mathematics with understanding and paying attention to pedagogical and psychological principles were mentioned in the report, and less drill and practice was emphasized.

Shortly after the war ended, the Board of Directors of the National Council of Teachers of Mathematics appointed a Commission on Post-War Plans to set direction for mathematics reform[45]. The first report, published in 1944, showed the importance of mathematics that had been demonstrated by the war—giving mathematics a positive image to the general public. The war showed that competence in mathematics could affect our survival. Therefore, the first report called for mathematical literacy to all who were capable. The Commission felt that the best way to improve mathematics overall would be the development of three tracks: one for lower ability related to consumer sciences; one for average ability involving mathematics taught for employment in modern business and industry; and another for college-bound students that would challenge and teach mathematics at a higher level than before. Furthermore, sequential courses had to be offered that were appropriate to changing needs. The Commission also stated its desire for improved teaching in all areas as well as guidance sources for students to enroll in appropriate mathematics classes.

The Second Report of The Commission on Post-War Plans, published in 1945, stated that the teacher must create experiences meaningful for the students to learn mathematics and then be able to apply what they had learned in other situations. The Commission strongly felt that mathematical reform needed to take place to reflect modern technological changes requiring citizens to have a higher level of mathematical understanding. The Commission stated that mathematics needed to be made applicable to real life as was happening in business, science, and industry. The war had taught many people that success in the traditional sequence of mathematics did not guarantee high mathematical ability. This second report also stated the need for a sequential mathematics curriculum that was challenging as well as somewhat uniform throughout our schools. Therefore, this report made specific suggestions for improving instruction in grades 1–4. The Commission stated, "as science and technology become increasingly important in the modern world, the functional uses of mathematics, likewise, become correspondingly urgent," and, "In fact, throughout the recorded period of human history, mathematics has been a mirror of civilization. It seems destined to continue in that role"[46,p.632].

The third report from the Commission, published in 1947, contained a checklist of twenty-nine specific topics that if students studied, they would have a solid competence in mathematics[47]. It was also known as the "Guidance Report" as it served as a guide for counselors and high school students. The following areas were discussed: 1) mathematics for personal use; 2) mathematics used in the professional workplace; 3) mathematics for the college-bound; 4) mathematics for the skilled workplace; 5) women in mathematics; and 6) mathematics used by government workers. This report, while stating the level of mathematics needed for specific careers, stressed that taking too few mathematics classes may not offer students a chance for a different career. Therefore, this Commission stressed that all students who were capable should take algebra, geometry, and trigonometry.

However, the most influential report or guidelines established during the "new math" were offered by The College Entrance Examination Board (CEEB). Concerning changing the mathematics curriculum specifically, The College Entrance Examination Board published a document in 1959 stating that mathematics instruction in public schools was generally in need of improvement as the level of mathematical competence of incoming freshmen in colleges was low. Students were not encouraged in mathematics, and schools had a very low percentage of students in high level mathematics courses—a fact that had become apparent after the launch of *Sputnik*.

The CEEB document first stated that the traditional requirements of one year of algebra and one year of geometry for entry to college were unsatisfactory. The CEEB document stated, "the handwriting is not only on the wall; it is all over the nation"[48,p.680]. This commission recommended a minimum of three years of mathematics for all college bound students and four years for those most capable. The most gifted and talented students should be placed in the Advanced Placement Program. Furthermore, the CEEB writers agreed with Conant that all mathematics classes should be based on ability grouping. CEEB also reported that while they believed calculus was a college-level course, gifted students may be exposed to it and analytic geometry through the Advanced Placement Program. In general, a push for challenging the academically talented began including an increase in the need for students to think critically, perform quantitative reasoning, deduce logically, and visualize spatially.

The CEEB report also stated that the fast-growing national need for people skilled in mathematics required an improved mathematics curriculum. The CEEB promoted the increased demand for people involving mathematics in industry, business, science, and many forms of employment. Furthermore, the CEEB mentioned the new computers that required mathematically able workers to operate, and the CEEB predicted a definite increase in the number of computers and workers that would be necessary in the near future.

The revolution in mathematics that began in the 1950s and continued into the 1960s was seen necessary because of three major things: increased mathematics research, automation, and computers[49]. Mathematics research had offered new mathematics material and topics such as abstract algebra, topology, functional analysis, and operations research. New content was changing the traditional mathematics high school curriculum and the undergraduate college curriculum. Automation in the form of mass telephone communication, airplane travel, guided missiles, and automatic industry machines was becoming a major part of American life. Therefore, students were expected to develop a higher level of mathematics consistent with the most technical world in which they were living. The third major cause was the development of the digital computing machines. Computers were making mathematics easier to apply in other fields as well as giving mathematicians the opportunity of working with formerly impossible computations very quickly and efficiently. Some content of the traditional mathematics curriculum was no longer necessary as new information and new automation made learning this content obsolete. This did not mean the elimination of key topics in mathematics, but instead students could dwell more on their application than their procedure. The time in school could be better spent on teaching newer content or applying mathematics in a different way than before.

Therefore, the "new mathematics" movement consisted of some new topics as well as traditional topics. But the biggest influences of "new mathematics" were not the "new" topics—they often looked like new concepts, but in reality old concepts were taught with new techniques. Innovations in language made old concepts appear new.

Teaching Standards

As America reached 1950, the increase in enrollment of students in schools caused an increase in the number of students taking mathematics and advanced mathematics courses[50]. Therefore, more mathematics teachers were needed, but the number of adequately prepared teachers of mathematics was small. In fact, there existed a shortage of mathematics teachers in our schools. By 1950, many high school mathematics teachers were teaching at a level beyond what they themselves had studied. Because of the continued shortage of mathematics teachers, many teachers of other subjects were being employed to teach mathematics despite their lack of ability in the subject. Therefore, by 1950, many students were discontent with the content of the mathematics classes and the manner in which material was being taught. The natural consequence was that students were being poorly taught, and negative attitudes about continued study in mathematics developed.

Moise stated that "in general the elementary teaching of mathematics has been a disaster"[51,p.1]. He believed that negative views of elementary teachers about the subject of mathematics had been passed to students. He also stated that in general, elementary teachers had taught nothing except arithmetic, and they had taught it only as a series of techniques or a means of calculation rather than a set of procedures. Therefore, something needed to be done not only about the mathematics curriculum, but also the standard of mathematics teachers at all levels.

While the first mathematics projects leaders were developing their new curricula and textbooks, the issue of pre-service training and in-service training became very important. In the 1950s, only about half of the states required a student to take mathematics to graduate, and many teacher training colleges did not require any mathematics to enter its program[52]. Even those students who had taken one year of mathematics to enter a teaching program often only had advanced arithmetic. Furthermore, the majority of the states did not require mathematics for elementary teacher certification. However, most teacher training institutions required elementary teachers to take one course in methods of teaching arithmetic but no mathematics content courses themselves[53]. Consequently, many elementary teachers certified to teach mathematics had the equivalent of a background in arithmetic. By the end of the 1950s, it was still estimated that over half of the high school mathematics teachers were inadequately prepared to teach higher level mathematics. Over 25% of high school teachers had never had a course in calculus themselves[54]. Therefore, teachers facing the "new math" materials had poor mathematics backgrounds themselves—with many elementary teachers having little more than arithmetic in their training.

In order to achieve these "new math" changes, the Commission on Post-War Plans recommended that elementary and secondary teachers have a strong, solid background in mathematics with continuous teacher education. The Commission also recommended that mathematics teachers have a major in mathematics, and even teachers in small schools who teach many subjects should have at least a minor in mathematics[55]. These recommendations would apply more easily to pre-service teachers provided the colleges and universities offered or required higher level courses in their pre-service teaching program. This would require colleges and universities to revise their teacher education and certification programs. However, the in-service training of current teachers of mathematics was seen as urgent.

There were various modes of re-education for these teachers. The most popular were institutes during the summer or regular year at a nearby college. The National Science Foundation also became prominent in offering summer workshops and training programs for teachers who needed to improve their own mathematics. The NSF paid the fees for these teachers to attend as well as paying the teachers to attend. Another option was for teachers to take a course called Contemporary Mathematics on the National Broadcasting Company's (NBC) Continental Classroom[56]. Teachers could watch this course on television from 6 to 6:30 a.m. during 1961-1962. This was the first time a course was offered on television for college credit. Many of the projects also began providing films for teachers to watch about how to teach with the new program. Teachers were extremely busy, and the retraining of teachers, while seen as necessary, could not totally happen in reality because of time constraints.

Pedagogical Techniques

Besides increasing the standard of mathematical content, teachers were asked to re-examine their teaching techniques. The traditional pedagogical techniques before the "new math" movements were those of the teacher as the instructor—lecturing while students learned passively. The traditional teaching had been criticized for requiring students to memorize facts and rules rather than truly understanding mathematics[57]. Repetition and constant drill and practice have been known as the standard teaching technique of mathematics.

Most of the "new math" project leaders advocated for "discovery learning" within their new curricula. Discovery learning is normally defined as a teaching technique where students are encouraged to learn mainly on their own to discover principles for themselves[58]. However, these project materials often utilized "guided discovery" or "discovery learning" similarly where the students participated more in their learning of mathematical concepts. UICSM, the first "new math" project, dwelled on two major teaching techniques that they defined as "discovery" and "freedom"[59]. In discovery, the students often discovered rules about numbers, and in freedom, students were given the freedom to attack a mathematical problem and answer it in their own way. Thus, students were asked to think and understand without necessarily doing the problem the way the teacher did.

Some guidelines for the teacher in the "discovery method" within the programs were associated with what we would most likely call "guided discovery" today. These included techniques such as the teacher questioning and patiently waiting rather than quickly supplying the answers. The teacher had to be able to follow the child's thoughts and support those that are correct and reject others that are not. Most importantly, the teacher could not quickly verbalize a general meaning before the student had understood and discovered a principle.

Discovery learning is an inductive approach of guiding students to understand mathematical ideas. The process has the following guidelines: 1) the main purpose is that students develop a pattern of thinking through the problem rather than searching only for an answer; 2) the teacher gives guidance where necessary while allowing the students to discover on their own when possible; 3) the students search for patterns, describe relationships they see, and form generalizations based on the mathematics problem they are facing, always taking an active role in learning; 4) modern textbooks provided questions to guide discovery learning for the students rather than only providing the answers; and 5) each topic is understood more when it is revisited later in the year or in other grades. Advantages to using the discovery method are a more permanent understanding is achieved, students develop a sense of freedom and confidence to learn, and classrooms are exciting.

Therefore, a teaching theory called constructivism became popular. Constructivists suggest that students learn more and better when they are allowed to construct meaning for themselves. They promote providing students with experiences where they can solve complex problems using skills that have been generalized beyond specific examples. In these cases, the teacher is often the facilitator of lessons. While constructivism is considered a philosophy of education, the teaching tool used is most often discovery learning. Many educators do not state that the learning of basic skills is not necessary, but they do agree that these skills do not have to be mastered before students can use them by applying skills to solve more meaningful problems[35].

However, it is important to note that while most "new math" project writers stated that they promoted "discovery learning," only some actually did so in their training and materials. Many project authors stated that discovery learning was good, but they definitely focused much more on content than pedagogy[60]. Furthermore, many of the "new math" programs that supported discovery learning were designed for college-bound students. One parent survey stated that guided discovery worked well for high ability students but can be very tiresome for students and teachers in a classroom of less interested or lower ability groups of students[61].

The traditional pedagogical role of the teacher prior to the "new math" movement was that of consultant and leader in education. This type of educational theory is often referred to as behaviorist, where the teacher has specific behaviors expected of each student while learning[62].   Students are expected to work systematically on mathematics, applying skills they have been taught to solve problems. Expected changes in behavior are easily seen as testing shows the increase in knowledge; however, the fault in this theory is that it does not allow for much abstraction[63].   Much of this traditional teaching method involved an emphasis on basic skills and drill and practice.

Many teachers have often felt that introducing mathematical topics too early would only harm children if they were not mature. However, during the "new math" movement, many psychologists stated that students could learn more than we have required in the past[64]. From the work by Bruner and Piaget, researchers in mathematics education began to appreciate the importance of psychology in education[65].

Piaget specifically stated that students could not learn formal operations or information too early because of their development. Piaget's learning theory can be described as follows: the child observes an event and forms a mental image, and the child revisits this mental image with continued observations until it fits into a pattern for him[66]. Therefore, teachers must provide students with experiences from which students may form mental pictures that are easily assimilated. According to Piaget, students who are not developmentally ready can not learn some higher level content because it is not easily assimilated. Piaget also believed that students could deal with problems when using concrete materials—therefore, discovery learning with manipulatives should benefit the student[67].   Some educators warn that because a student can learn more does not necessarily mean he or she should learn more.   Young children should be allowed to explore mathematics with less formalization.   Piaget who had been writing about his work for years, did not become famous in mathematics education until the 1960s—especially for the developmental stages he identified[68].

Bruner further supported the idea that simple concepts are further developed with the addition of new concepts[69].   Bruner promoted his ideas with the spiral curriculum where students would continually learn more about a topic at different times when building on what they already knew. Bruner felt the way to achieve this learning was through the discovery method.

The "spiral curriculum" is mentioned throughout the "new math" materials. The spiral curriculum is defined as a curriculum in which concepts are first introduced at an early stage and then returned to again and again each time at a higher level than the time before[70]. Therefore, the students will eventually have a high level and much deeper understanding of a topic.

In addition to considering specific teaching techniques or developmental stages, a number of researchers emphasized that mathematics teachers must teach with not only a solid background in mathematics, but also with enthusiasm[71]. Mathematics for children is intimately influenced by the teacher's attitude[72]. This can become a negative cycle if teachers with negative attitudes teach children who develop negative attitudes, and these children become adults influencing more children with negative attitudes. Hostility towards mathematics is most often seen in elementary teachers even during their pre-service education; therefore, fostering more positive attitudes during their undergraduate career would possibly help break this cycle.

Requiring unqualified teachers to teach mathematics can often result in a negative attitude displayed for the students which may encourage the abandonment of further mathematical study. Teachers must make the subject appealing and interesting to students. Teachers must be personally interested in mathematics and encourage all students while finding and challenging talented students. Thus, the issue of motivation and positive mathematical attitudes fostered in students is equally crucial to strong content.

> I must emphasize that the elementary school teacher, the junior high school teacher, and the senior high school teacher are absolutely essential to the success of our program to provide better mathematics; for these teachers must teach mathematics; and these teachers must teach with enthusiasm so that their students continue the study of mathematics[73, p.14].

An important issue in mathematics should be encouraging students to develop problem solving skills. Problem solving and critical thinking are essential skills for students not only in mathematics, but in life; therefore, both should be taught and incorporated in mathematics classes. In fact, critical thinking would become one of the main goals of the "new mathematics" programs as the materials would be designed for students to think, reason, and deduce.

Another technique that teachers began to use more was the introduction of manipulatives which are described as materials that students can normally use to explore in a "hands on" manner in classrooms—often making understanding a difficult concept easier. Manipulatives allowed for more guided discovery learning as students explored mathematics. Teachers began abandoning conventional teaching techniques such as drill and practice and mathematics taught as a sequence of manipulations. Instead, they began using different teaching techniques under a different teaching philosophy during the beginning of the "new math" era. The use of guided discovery with manipulatives and a stress on problem solving was encouraged so that students could truly *understand* mathematics rather than simply solve problems with little meaning.

"New Mathematics" Projects

World War II underscored the educational needs of a technological economy. After the war the requirements of industry in engineering, the sciences, statistics, and mathematics expanded dramatically. The gap between the mathematics taught in high schools and the mathematics required for postcollege jobs grew wider year by year, for industry (and national-defense research) required ever more sophisticated mathematical techniques. Something had to be done[1, p.23].

Some projects were developed in the 1950s due to pressure to increase the mathematical ability of college-bound students. However, with the release of significant amounts of federal funds in the United States after the launching of *Sputnik*, a wave of reform projects developed, and these ideas affected the United States and then crossed the ocean to England and Western Europe[2]. By the early 1960s, many similar textbooks began to appear with this "new" curriculum developed by many different reform projects. By 1962, the revolution in mathematics was in "full swing". During the "new mathematics" era, there were over 600 groups, programs, or projects—some of which put teaching materials together quickly[3]. The "new mathematics" projects were, for the most part, positively presented and received, as can be seen in popular magazines from 1956 – 1964[4].

Chapter Three contains descriptions of many of the major mathematics projects of the "new mathematics" era; most funded as a result of the launching of *Sputnik*. Various mathematical or curricular organizations that promoted "new mathematics" will also be discussed. The following is a list of the major projects, the minor projects, and curriculum groups with their starting dates.

Major Projects
- University of Illinois Committee on School Mathematics (1951)
    - University of Illinois Arithmetic Project (1958)
- University of Maryland Mathematics Project (1957)
- Commission on Mathematics of the College Entrance Examination Board (1959)
- School Mathematics Study Group (1958)
- Greater Cleveland Mathematics Project (1959)
- Madison Project (1957)
- Comprehensive School Mathematics Project (1963)

Minor Projects
- Developmental Project at SIU (1958)
- Boston College Mathematics Institute (1957)
- SMP (1962)
- Nuffield Project (1964)
- Ontario Math Commission (1959)
- Minnesota Mathematics and Science Teaching Project (1961)
- Ball State Teachers College Experimental Project (1955)
- The Suppes Project (1958)
- Other state and school projects

Curriculum Groups
- Cambridge Conference on School Mathematics (1963 and 1966)
- Secondary School Mathematics Curriculum Improvement Study (1965)
- National Association of Secondary School Principals (1958)
- National Council of Teachers of Mathematics (SSCC report in 1958)

After explaining the various projects and curriculum organizations, the author will discuss the similarities among most projects, contributing to issues pertaining to the entire "new mathematics" movement.

Major Projects

While many factors contributed to the "new math" movement of the 1950s and 1960s, major projects lead this reform as leaders and models for other projects. Many of the major projects received significant funding from both private sources and from the government. Most major projects were developed in distinguished universities by prominent leaders in mathematics and mathematics education. Most developed new curricular materials to promote a curriculum focused on the understanding of mathematics. An analysis of each major project is provided.

University of Illinois Committee on School Mathematics

The first of these large-scale projects to deal with school mathematics was the University of Illinois Committee on School Mathematics (UICSM) which began in December 1951[5]. The original name for this project was the University of Illinois Committee on Secondary School Mathematics[6]. This project's writers began to investigate mathematics content and teaching in high school under the direction of Max Beberman. The major concern was the development of materials and their experimental trial in schools throughout the country. UICSM was formed because the University of Illinois desired a change in the mathematics standard of its incoming freshmen into the College of Engineering[7]. The UICSM project received funding from the University of Illinois, the Carnegie Corporation, the U.S. Office of Education, and the National Science Foundation[8].

Traditionally, undergraduates entered the College of Engineering with little mathematics, requiring the university to teach calculus their sophomore year—thus delaying the introduction of physics in the curriculum[9]. The College of Engineering wanted all freshmen to take analytic geometry and calculus their first year of college—this required moving trigonometry, algebra, and some analytic geometry into high school from its traditional teaching place in college[10]. Therefore, UICSM started its own classes in its laboratory high school which resulted in some calculus being taught in twelfth grade. This step made UICSM a leader in the introduction of calculus as a high school course. Therefore, UICSM was established because of the need for schools throughout Illinois, particularly the small schools, to prepare students with high enough standards that they could enter the College of Engineering at the University of Illinois[11]. While UICSM began its project with the idea of improving prospective engineering students, it soon realized that focus should be spread to include all capable students as they would benefit as much from this new program as the prospective engineering students.

UICSM felt that much improvement could be made to the secondary mathematics education curriculum across the country in all public schools. UICSM also reported that in general, America was lagging in producing students with strong mathematics and science backgrounds, especially college-bound students. Therefore, the University of Illinois issued a bulletin in 1951 describing new entrance requirements for college engineering freshman—particularly advanced mathematics classes. This bulletin was written jointly with the colleges of education, engineering, and the department of mathematics, and was titled *Mathematical Needs of Prospective Students in the College of Engineering of the University of Illinois*[12].

The bulletin explained that students should have completed all elementary algebra, geometry (both solid and plane), and trigonometry before entering college[13]. The bulletin also listed all topics that students should have covered in their high school classes to be prepared to begin analytic geometry and calculus their freshman year of college. The bulletin stressed that this list of requirements was the minimum for pre-engineering students, and students who could take higher levels of mathematics in high schools were encouraged to do so. Students who had not completed these requirements may be allowed entry to the program, but would be required to take these courses which delayed their graduation by at least one semester. The University of Illinois Laboratory High School joined these forces in making the UICSM project a reality. The use of this laboratory school gave the University of Illinois the means to develop a new secondary school mathematics curriculum. The UICSM committee was very proud of the fact that a revised mathematics curriculum was being developed with successful collaboration between teachers, mathematicians, and mathematics educators[14].

UICSM felt that the traditional role of algebra in ninth grade, geometry in tenth grade, another course of algebra or geometry in eleventh grade, and trigonometry in twelfth grade need not be necessary in that order. UICSM suggested that mathematics be taught together incorporating many ideas rather than subject by subject. Therefore, the course materials and textbooks developed by UICSM incorporated different topics throughout each year of schooling, and essentially combined traditional college-preparatory four year programs into three years—partially by eliminating repetition[15]. This kind of curriculum is often labeled unified mathematics or integrated mathematics. UICSM adopted a spiral system of integrating some information from arithmetic, algebra, and geometry together in each year's course. An advantage to this type of integrated program was that students could easily see the connections among different mathematics principles. The UICSM outlined its first new course in the spring of 1952, and their experimental high school was using the material beginning fall of 1952[16].

Another principle that UICSM held was that mathematics needed to be made real to the students in their present world of ideas and abstraction. While UICSM stated that drill and practice was necessary to some extent, the teacher must make mathematics meaningful for the students so that they truly understand it. Once the students understood, many of the homework exercises were drill and practice[17]. The ideas pertaining to the content of this college-bound program were similar to those expressed by the NCTM Commission on Post-War Plans when describing the third track of mathematics that should be offered to capable students. Therefore, UICSM did not begin their reform with the idea of radically revising mathematics—instead, they were concerned with improving the teaching of the traditional curriculum and making the transition from school to college easier.

However, as UICSM began involving research mathematicians, new ideas were incorporated in their curriculum. This also marked a difference in mathematical reform—the influence of research mathematicians had never been considered in revising school mathematics, and UICSM was the first project to do this. Texts were developed incorporating new ideas with traditional content, and teachers were instructed in the use of the materials in summer training institutes in both content and pedagogy. UICSM made units for grades 9–12, and this four year high school curriculum consisted of arithmetic, generalizations and algebraic manipulations, equations and inequations, ordered pairs and graphs, functions, geometry, mathematical induction, sequences, elementary functions, trigonometry, and polynomial functions[18]. The "new" sections of mathematics education that UICSM added were based on material of interest to research mathematicians that was considered interesting for high school students. The "new sections" mainly consisted of all topics listed above with the exception of traditional arithmetic, geometry, algebra, and trigonometry. UICSM developed its curriculum for college bound high school students, but modified it to address curriculum issues for high school students of all abilities and finally for elementary students. UICSM promoted a one-track mathematics curriculum that had extremely high standards. The idea set forth was that the college-bound would take four years of this mathematics curriculum while the non college-bound could take two or three years—a course still offering high mathematics ability that UICSM felt was needed in all students[19]. However, the first courses were developed especially for college-bound students.

Beberman, the leader of UICSM, stressed that the most important issue in the new curriculum was not the content, but the fact that students must really understand what they were learning. Beberman also expressed that students can only truly understand mathematics when they have the chance to "discover" principles on their own with guidance by a teacher using unambiguous and precise language. The student was expected to form abstractions and solve mathematics using problem-solving techniques. Logic was also emphasized in the UICSM curriculum. UICSM was a promoter in the distinction between numeral (name given for a number) and number, between number and variable, and to stress that equations are sentences using precise language[20]. Therefore, discovery and precision in language were the two key elements in promoting understanding. Another emphasis of the UICSM project was consistency in language and explanations as well as structure of mathematics[21]. The importance of mathematics structure was described as the ability of students to appreciate that the number systems used are one kind of an entire mathematical system[22]. UICSM felt that by following these principles, mathematics could be enjoyable for students because students have an interest in ideas and creativity.

Discovery is described by Max Beberman in the following way: "To us this means that after we have selected a body of subject matter to be learned we must design both exposition and exercises in such a way that the student will discover principles and rules"[23,p.24]. His definition sounds much like what we describe today as guided discovery; the teacher has input in "exposition and exercises" as seen above. Guided discovery is often defined as a teaching method in which the teacher or schools sets goals and content, but the student develops procedures and rules along the way. Thus, when discussing the "new math" projects, mention of the word discovery can be assumed to mean "guided discovery as our definition exists today—unless otherwise indicated. Beberman stated that some believe true discovery learning without guidance from the teacher may be best, but UICSM stated that the use of guided discovery techniques are better for a curriculum developed for a wide variety of schools and teachers. Beberman admitted that guided discovery will take more time than traditionally taught material; however, the benefit of students understanding and developing a strong proficiency in mathematics is worth the extra class time that is often used for "drill and practice." Beberman also stated that the two most important aspects in mathematics to maintain interest for students are sensible problems and opportunities to solve problems differently. Therefore, the benefit of discovery learning is not in the ability to solve real-life problems, but is instead in developing interest in mathematics and increasing mathematical reasoning and thinking.

Initially, UICSM only allowed the use of their materials by teachers who had been instructed in their institutes, and they often supervised teachers in cooperating schools. Teachers were expected to participate in an orientation program at the University of Illinois or in another summer institute elsewhere; they were required to have one extra free period each week in order for the extra preparation that was needed to teach the UICSM materials; and they were expected to follow the UICSM guidelines and report weekly to the project center[24]. In 1958, the texts became available for more general use. UICSM continued to promote their materials only with well educated teachers who had access to the supplementary guides, and they also sent a newsletter to all teachers using the supplies to keep the teachers abreast of new research or ideas. UICSM also made films that could be watched by teachers or other educators interested in seeing how to incorporate discovery learning and use these new materials[25]. These films were to be used separately or in conjunction with teacher training institutes, and were tested during 1960[26].

UICSM stressed discovery teaching in the use of its materials, and the units they had developed were to be followed sequentially. The discovery method was built into the UICSM materials so that students could continue to discover while completing homework. The differences between the UICSM curriculum and the traditional normal curriculum were the following: 1) units moved toward an integrated curriculum with algebra throughout; 2) geometry was minimized from the traditional two years; 3) trigonometry was not a separate course; 4) modern terminology was used throughout; and 5) some mathematics topics were taught earlier to younger students[27]. Also, the program attempted to align its approach with contemporary mathematics; to encourage creativity; to present mathematics as an integrated subject; and to develop problem solving skills by providing students with challenging and new problems[28].

The six main principles of UICSM were: 1) consistent presentation of high school mathematics can be developed; 2) maintain student interest in mathematics; 3) manipulatives should be used in teaching mathematics; 4) language should be unambiguous; 5) student discovery should be encouraged; and 6) students must understand mathematics. In fact, "discovery teaching and learning became the hallmark of the UICSM program"[29, p.254].

UICSM studies showed that high ability students using their materials made a significantly greater gain in understanding basic mathematics concepts compared to a traditional class[30]. However, UICSM also promoted increased mathematics for all students—not only college bound students. UICSM saw its program as an attempt to improve student ability and attitude for all schools using its program. UICSM began development of programs for low-ability students in the early 1960s. However, most of UICSM's work was completed within ten years of its beginning (by 1962). And by 1960, approximately 12,000 students from schools affiliated with the UICSM project were using UISCM materials[31]. As this number of students accounted for only one-third of all materials sold by the University Press, it was assumed that many other schools were offering the course.

UICSM limited the widespread use of its program by the requirements placed upon those involved. Teachers spent much time in training and sent weekly reports and continual correspondence to the program. Many schools wishing to implement the UICSM program were encouraged to have at least two teachers in the school using the program so that they could use each other as resources. Members of the UICSM staff frequently attended classroom sessions in various schools. By 1958, fifty-five schools in approximately twenty different states were using UICSM materials[32]. The leaders of UICSM proceeded with caution in order to ensure that their materials were being used as they had planned. They also invited teacher associates from school districts or other universities to participate on-site at the University of Illinois for one year. While on leave from their principle position, they became involved by teaching in pilot schools, participating in staff seminars, corresponding with participating teachers, and visiting pilot schools. Their salaries were paid partly by

UICSM and partly by their sponsoring institution. UICSM admitted that widespread use of this program would be unlikely due to the nature of the program.

UICSM was credited with continually revising materials and techniques to improve the program. Even though the UICSM materials were not in widespread use, the value of the UICSM project was in the example and working model it provided for the development of other "new math" projects. Creating change in mathematics education was one of UICSM's greatest accomplishments. UICSM's format of creating new curriculum materials, training teachers to use these materials, and testing the use of the materials in cooperating schools established precedents that became a pattern for later projects[33].

By the early 1960s, Beberman saw the desire to create more materials based on what the committee had learned since the beginning of testing their new curriculum. He decided to focus on grades 7–12 rather than only high school[34]. Beberman felt that much of what was taught in junior high mathematics traditionally was a waste, and the valuable high school mathematics could be pushed down somewhat in order to allow more time in high school to pursue traditional and modern ideas. Beberman continued to stress the importance of testing materials developed by educators and mathematicians because he did not know what would be best for the children without trying these new ideas.

UICSM saw a need for the development of materials for ages other than secondary school students. Therefore, another group spawned from UICSM called the *University of Illinois Arithmetic Project*. David Page became the leader of this elementary education reform by UICSM. The University of Illinois Arithmetic Project began in 1958, and was designed to develop materials for grades 1–6[35]. The purpose of this group was to find more and better content for elementary mathematics. This group believed that elementary school children could learn much more mathematics than they were being taught—particularly because students naturally want to learn more. The leaders of this project felt that pressure to wait for more development among children before teaching higher levels of mathematics had hindered the child's ability to learn the language of mathematics." The project writers also felt that students should learn at least as many arithmetic skills as had been taught previously[36]. Students using the project materials were found to have equally solid computational skills while a great deal of new mathematics was taught with the guided discovery method.

The project planned its curricular materials around the idea of guided discovery, as had been the case for the UICSM project. Topics for elementary grades included linear equations, positive and negative numbers, metric and English units, probability, geometry, and mental arithmetic. The project did not intend to develop full curricula for all grades, and instead, concentrated on supplemental learning and teaching traditional concepts better[37]. The Arithmetic Project hoped to first address the teaching of teachers in this new material, and the results would then gradually affect elementary students. In fact, one of the main goals of the Arithmetic Project was to train teachers to be able to instruct with this new material in a way that students became excited about mathematics[38]. The project did this by encouraging teachers to observe classes and work similar problems. Teachers were encouraged to begin topics with their classes after a few weeks of training, and many reported that they saw an immediate change in their students' motivation, interest, and involvement in mathematics.

The Arithmetic Project was based on the following principles: 1) children of all ages can learn and enjoy learning; 2) children learn about interesting topics through guided discovery as well as computational skill; 3) teachers must have a solid background in mathematics, but they must also be taught mathematics with a positive attitude; 4) teachers only see the benefit when they see their students excited and learning[39].

The essential elements behind the projects at the University of Illinois to improve mathematics education were sound mathematics and meaningful pedagogy. The curriculum hoped to replace the conventional curriculum with more geometry and integration of science with mathematics in order that students understand mathematics and enjoy it.

University of Maryland Mathematics Project

The University of Maryland Mathematics Project (UMMaP), also known as the Maryland Project, began structuring its new mathematics curriculum in the fall of 1957[40]. It was under the direction of Dr. John R. Mayor as a project to improve mathematics in grades 7 and 8. Mayor felt that much time is wasted in mathematics during junior high school as many highly capable children are bored and many less capable children are tired and frustrated with drill and practice.

This project was started with cooperation between the departments of education, mathematics, engineering, and psychology at the University of Maryland[41]. Funding was available from the Carnegie Corporation and the university[42]. They, like UICSM, encouraged close contact with participating schools. The four major public school systems in the Washington area participated, and the teacher-writers for UMMaP came from these districts.

UMMaP followed the lead of UICSM by emphasizing psychological research relating to mathematics education, particularly in the role of discovery learning. UMMaP's psychological research in mathematics education was superior to that of UICSM. For example, the purpose of UMMaP was to determine the maturity levels of children at which specific mathematical topics could be learned appropriately[43]. The applications from psychology used in UMMaP were from Gagne's learning theory which states that children must learn new things only after already learning the basis upon which to build[44]. The materials must be psychologically sound, must be strong in content, and must be applicable to modern needs. UMMaP focused on the following four issues: 1) simple but precise use of language; 2) emphasis on understanding rather than rote learning; 3) integrated arithmetic and algebra; and 4) mathematics structure[45].

The UMMaP began its curriculum foundation with production of materials for grades seven and eight. UMMaP presented all of the traditional ninth grade algebra content into their seventh and eighth grade texts. Junior high topics also included logic, the solution of triangles with trigonometry, irrational numbers, graphing, statistics, and probability[46]. A text with a teacher manual was available in February 1961. The importance of mathematical language was stressed. The difference between number and numeral was stressed as was the case with the UICSM materials[47]. The goal of UMMaP was to develop a junior high school curriculum that provided new content as well as greater understanding of mathematical concepts[48]. Emphasis was on understanding as opposed to rote learning or arithmetic skills.

By 1960, approximately 15,000 students were using the UMMaP materials[49]. Some early evaluations comparing students studying the traditional curriculum and students studying the UMMaP curriculum showed that students studying the UMMaP curriculum performed as well on traditional tests and much better on "new mathematics" tests as students studying the traditional curriculum[50]. However, this appeared to be true mainly for average and above-average students. Those lower ability students seemed to struggle with the new curriculum and performed worse on traditional tests than if they would have taken the traditional course[51].

UMMaP has been regarded as one of the few serious attempts to create a modern mathematics program for the junior high level. Furthermore, its materials were highly regarded because UMMaP had tested its own materials throughout development. The materials were initially developed to bridge the gap between arithmetic and high school mathematics. One of the greatest feats of this project was the proof that course materials could be developed successfully with the cooperation of the mathematician, the classroom teacher, the supervisor, and the administrator. However, like UICSM, UMMaP had not spread far in the US before larger projects came about. Research shows that the UMMaP and SMSG (to be discussed shortly) projects developed almost identical junior high textbooks[52].

The Commission on Mathematics of the College Entrance Examination Board

The Commission on Mathematics of the College Entrance Examination Board (CEEB) was a group that affected mathematics reform by utilizing high school teachers, college mathematics professors, and college mathematics education professors to develop a new curriculum[53]. The CEEB had been formed in 1900 to provide examinations that would help standardized college entrance requirements. Therefore, CEEB began working with secondary schools prior to 1950 to develop an Advanced Placement (AP) program for the mathematically gifted[54]. This specific group hoped to recognize the mathematically gifted and develop their strengths to the highest potential in high school and college. The first AP examinations were written in the spring of 1954, and they were readily accepted by many colleges[55]. The Advanced Placement program allows for students to take examinations covering this high level of material, and if they succeed, they are given college credit and placement into a higher level mathematics class at college.

The Commission of the CEEB completed an investigation of the mathematics curricula in US secondary schools and made recommendations upon its findings. The CEEB stated that mathematics was changing and that changes should be reflected in our country's high schools. This commission had a desire to "replace present 'seventeenth-century' programs in the schools"[56,p.1]. New mathematics topics were exciting and needed to be included in the traditional curriculum. However, the CEEB felt that new topics needed to be introduced in terms familiar to students and educators such as within courses like algebra, trigonometry, etc. These new topics included probability and statistics, topology, and modern abstract algebra. Therefore, they were the first group who explicitly promoted the idea of presenting new subject fields into the curriculum rather than changing pedagogy or the presentation of traditional topics. This curriculum was designed for college bound students and was based on concepts and skills, deductive reasoning, mathematics structure, sets and functions, and reorganization of geometry, trigonometry, and final year mathematics. The purpose of the Commission of the CEEB was,

> To consider broadly the college preparatory mathematics curriculum in the secondary school and make recommendations looking toward its modernization, modification, and improvement. The overriding objective of the Commission has been to produce a curriculum suitable for students and oriented to the needs of mathematics, natural science, social science, business, technology, and industry in the second half of the 20th century[57, p.28].

While the curriculum was set forth for college-bound students, the CEEB did state in its report that some mathematics should be studied by all. More specifically, all students should have an understanding in arithmetic, elementary algebra, general properties of geometry, deductive reasoning, and value of mathematics. The curriculum was developed into a proposal called, the *Program for College Preparatory Mathematics*, and while it was not published until 1959, its contents were widely known and circulated before then. Some of the influential people involved were Max Beberman of the UICSM program, Edward G. Begle of the SMSG program, G. Baley Price of the MAA (Mathematical Association of America), and Frank B. Allen of NCTM.

In this proposal, the CEEB called for the implementation of new mathematics content, the reorganization of older mathematics, and newer uses of mathematics. The CEEB did not intend to uproot tradition; but they did want to improve mathematics education based on current research. Older mathematics should not be forgotten or deemed as less worthy because of newer mathematics brought into the curriculum. "The permanence of traditional mathematical ideas must not be forgotten"[58,p.670]. However, mathematics was increasingly needed for all people as its applications are apparent in science, social science, business, industry, and other fields of applications. Furthermore, with the electronic computer and data-processing systems, mathematics could now be used as a means of organizing masses of data and information not possible in the past. Some traditional mathematics may have seemed less important since "newer" topics could be understood without the same traditional basis. Therefore, one goal was to find a balance between traditional topics and new topics while incorporating more students with higher ability overall in mathematics.

The CEEB did mention that teachers and textbook writers should develop programs with understanding of how children learn, but they did not offer any guidance or suggestions concerning pedagogy. The only comment they did make was that they felt students should discover as much as possible while learning mathematics, but they did concede that the discovery method needed more time in the classroom which may not be available. In general, the CEEB's proposal was most concerned with content. However, the CEEB did stress that good teaching of mathematics would be vital to success of this program[59].

The CEEB stated that despite the form of school organization (often referred to in terms of 8-4, 6-3-3, 6-6, etc.—indicating the number of years in elementary and secondary schools respectively), topics for grades 7 and 8 should consist of the following: arithmetic, geometry, algebra and statistics. The CEEB recommended that ninth grade mathematics mainly consist of algebra—reasoning and manipulative skills taught at a higher level than traditionally. They recommended solid and plane geometry be placed together for grade ten, and that the number of theorems be reduced while including some coordinate geometry. The CEEB recommended that trigonometry, infinite series, and vectors be taught in the third year of high school. The CEEB hoped that deductive thought could be used throughout mathematics rather than prominent only in geometric proofs[60]. The twelfth grade recommendations were listed separately because the CEEB believed that only students pursuing mathematically related careers in college would proceed to the fourth year of mathematics in high school. They recommended that sets and functions be taught in the first semester of fourth year mathematics and that the second semester contain either abstract algebra or probability and statistics or selected topics. The CEEB, however, did not recommend that calculus be taught in high school—they saw this subject for college level students. The CEEB strongly emphasized the study of probability and statistics because they saw statistical inference as extremely important for the future of all students despite their specialty. The CEEB called for change because of the following criticisms of the traditional program: 1) too much emphasis on manipulation in algebra; 2) deductive reasoning ignored except in geometry; 3) rote memorization of theorems in geometry; 4) too many topics that had become obsolete; and 5) the need for incorporating new topics[61].

The CEEB proposal warned of too much drill and practice and emphasized teaching for understanding. They stated that mathematics is too often taught as a "bag of tricks" rather than a creative process. They felt that they were offering little in the form of "new" content but instead, they were emphasizing important changes in instruction and content emphasis. In reality, the CEEB proposal was most in line with the goals of the SMSG project (to be discussed shortly)—dwelling much more on content than pedagogy. However, the CEEB proposal did state that the major demands of any curricular revision were placed in the hands of the teachers. The CEEB was recommending a program that gradually grew out of the old in a manner that was reasonable for teachers to adjust. The CEEB was interested in making their report widely known, and the original printing allowed 39,000 copies to be sent to school administrators and other interested and influential educators[62]. The CEEB then stated that they would begin changing their examinations to match the curriculum once it had been altered in the nation's high schools. The CEEB did not produce any curricular materials—they only made recommendations for changes.

One of the largest differences and effects of the CEEB's work is that for the first time in this era of reform, a nationwide improvement was called for in mathematics in all schools. Widespread distribution to all parts of the country caught the attention of mathematics reform nationwide. "With the Commission's attempt to sell curriculum reform on a nationwide scale, the 'new math' moved from the experimental stage to the evangelical stage"[63, p.4].

School Mathematics Study Group

The School Mathematics Study Group (SMSG) formed in 1958 as the product of public concern and high school administration expectations to "do something about mathematics" after the launch of

Sputnik. The primary purpose of SMSG was to foster research and development in the teaching of mathematics in schools[64]. SMSG hoped to provide an improved curriculum by which students would learn basic skills as well as a deeper understanding of mathematics. The goals of SMSG were stated as 1) to provide college-bound students (average and above average ability students) with a mathematics program in high school that would enable them to take calculus the first year of college, and 2) to provide the less capable student a mathematics curriculum that the student was capable of understanding the nature of mathematics and the role of mathematics in our society[65]. However, many felt that SMSG was formed as a "sponsored crash program" to alleviate the scientific problems of the United States.

SMSG formed from two major conferences: one NSF conference held at the University of Chicago and another NSF conference held in Cambridge, Massachusetts. The National Science Foundation had sponsored a conference on the production of research mathematicians at the University of Chicago in February 1958 where the American Mathematical Society appointed a committee to deal with mathematics pre-college problems[66]. Another conference, also scheduled by the National Science Foundation, was held a week later in Cambridge, Massachusetts, where a physics committee was already working on a more modern curriculum. This committee included physicists, educators, and high school physics teachers working together to create new text materials for high school physics. The sole purpose of this meeting, which became known as the Cambridge Conference, was to discuss the future of secondary school mathematics in a place where mathematicians could consult with physicists who had been working on revising their curriculum for over two years.

Because of the content of both these meetings, the School Mathematics Study Group was named in consultation with the presidents of the National Council of Teachers of Mathematics and the Mathematics Association of America and the American Mathematical Society to focus on the improvement of the teaching of mathematics in schools[67]. The involvement and collaboration of the presidents from these organizations was the first one of its kind concerning elementary and secondary schools. Furthermore, the establishment of this committee also supported collaboration among mathematicians and teachers of mathematics.

This group was led by Edward G. Begle at Yale University, and they began with the intention of concentrating on creating a new 7–

12 curriculum. Once this curriculum was initially established, SMSG saw the need to consider developing a "new mathematics" program for the elementary grades. The members of the group included university mathematics professors, high school mathematics teachers, mathematics education professors, and representatives of science and technology. The aim of this group was to improve the teaching of mathematics, to persuade more students to study mathematics, and to ensure that mathematics content was appropriate for today[68]. The group hoped to do this by teaching for understanding and insight without neglecting basic skills. The National Science Foundation (NSF) funded much of the group's research. The popularity of SMSG developed rapidly because they could achieve quick results through the production of textbooks. Since the public, press, and the government wanted immediate results in improved mathematics, many supported the initial work of the SMSG.

During the first writing session, SMSG analyzed methods and philosophies of the physics project, the UICSM project, and the UMMaP project[69]. In fact, some of the writers for SMSG were also members of these other projects. SMSG used the CEEB's report when it began in 1958 to create new content material. SMSG focused on building a mathematics curriculum that used basic and advanced mathematics skills and attracted mathematically able students. SMSG considered college-capable students as the upper third of each class. SMSG also stated that they stressed training teachers adequately to teach this new curriculum; however, many would disagree that this happened. The basic philosophy of SMSG was that a more advanced level of mathematics should be introduced earlier in the mathematics sequence from elementary to high school. This allowed for more advanced classes to be taught at the high school level than ever before. New materials were developed to accomplish this goal. The first versions of the materials were available in 1959, and after one year of trial, they were available in their revised form for use in schools in 1960[70].

Concerning the junior high level materials, SMSG stressed recognizing patterns, mathematical vocabulary, deduction and induction, some informal geometry, measurement, and some probability and statistics. SMSG found that they could borrow much from UMMaP, and the writers worked in close cooperation with the University of Maryland[71]. The algebraic and geometric structure of mathematics was emphasized in the high school materials. The high school curriculum was revised to incorporate algebra in ninth grade, geometry in tenth grade, and advanced algebra and trigonometry in eleventh grade. This restructuring of topics allowed for all four years of the traditional mathematics curriculum to be taught in three years— leaving the fourth year of the high school mathematics curriculum to be developed by the projects. Because SMSG felt that calculus was really a college-level course (particularly because of the lack of qualified teachers to teach calculus at the high school level), SMSG developed their twelfth grade curriculum to include abstract algebra and the study of functions. SMSG hoped to eventually have approximately six different courses for the twelfth grade. As can be seen by the above outline of topics, SMSG followed the ideas established by the UMMaP and UICSM groups, but their curriculum very closely followed the CEEB report which was published after the first SMSG writing session but of which preliminary drafts were available.

SMSG realized the importance of educating the teachers, but did this mainly through teacher manuals. SMSG suggested teacher in-service programs and more time allocated to teachers who were using SMSG materials for the first time to understand the structure and content of the program. SMSG had three goals: 1) develop both an improved basic and advanced curriculum; 2) attract and train capable students in mathematics; and 3) provide assistance to teachers who were going to use SMSG materials[72]. While the initial goal of SMSG was to develop materials for high school college-bound students, they were aware of the need to develop materials for all mathematical abilities as well as elementary grades. SMSG believed that elementary students should be taught mathematics with an increased emphasis in some newer topics such as geometry as well as an overall emphasis in mathematical understanding.

The groups did this by eventually developing sample textbooks for all grades K–12; providing teacher manuals and enrichment materials; and by testing the program materials. SMSG wanted its textbooks to serve as guides, but originally hoped that they would be in use only until more publishers could provide other books. SMSG also stressed the discovery method in its teaching materials. However, SMSG promoted limited pedagogical efforts because this would take longer than developing the content in textbooks. For example, researching teaching methods was considered more difficult than researching the testing abilities of students using SMSG materials. Furthermore, mathematicians dominated the SMSG efforts resulting in a modern mathematics program based on strong content with little pedagogy which was one of the primary purposes of the first "new math" projects. This program reflected the curriculum of a new program developed quickly for rapid change as wanted by the government.

Testing of the original SMSG materials began in "centers" chosen by SMSG where a consultant worked with willing teachers to try this new curriculum. These mathematical consultants were chosen jointly by SMSG and the school district. SMSG, funded mainly by the National Science Foundation, paid these teachers to try their new materials and also paid these mathematician consultants to work part time in consultation with participating schools who were also provided with the textbooks at no charge[73]. SMSG was the first "new math" project primarily funded by the NSF as UICSM, CEEB, and UMMaP were mainly funded by the Carnegie Corporation. Textbooks were given only to these testing centers until 1960 when they were made public[74].

The elementary program established by SMSG was criticized for its unrelated or non-connected topics as well as for its use of advanced symbols and "set" language. The high school program was criticized for its unusual manner in presenting traditional topics including introducing too many topics at one time. Furthermore, SMSG had different writing teams for each grade level that did not necessarily collaborate with each other—thus some duplication of topics existed. Both programs used traditional pedagogical methods with some discovery teaching, and precision in language was emphasized. While SMSG said they promoted discovery learning, their materials did not present content in a discovery method, but instead, presented it by means of exposition. However, the precision in language was superb by first showing examples of new concepts, then making generalizations followed by deductive proof. One major complaint in the language use was that "sets occur everywhere!"[75,p.13].

SMSG produced many research projects about its program, particularly the SMSG National Longitudinal Study of Mathematical Abilities[76]. Research showed that pupils who studied with SMSG learned traditional material as well as new material, thus reinforcing the hypothesis that high school students are capable of learning college level mathematics. Furthermore, high school students studying with SMSG materials performed as well as traditional high school students on standardized tests of traditional material and scored high on ability to understand mathematics. Many students and teachers who used the new materials also gave favorable feedback to SMSG. While many parents supported the work of SMSG, the biggest complaint was that they felt incapable of helping their children with homework.

The effects of SMSG were far greater than the gains of the students. SMSG is known as the largest national project developed to improve the mathematics curriculum[77]. During the 1959-1960 school year (the first year SMSG used its materials), it is estimated that 26,000 students were using SMSG materials[78]. Therefore, SMSG was influencing teachers and students at a much faster rate than any other reform project. Furthermore, SMSG provided materials enabling any class to begin at their appropriate age level without having studied SMSG material previously.

The writing sessions showed that college and high school faculty could work together successfully. The SMSG project was used as a model for the development of later projects. The popularity of SMSG can be attributed to the speed in which it researched and developed new textbooks, and the government source "advertising" its immediate implementation. Another factor contributing to the popularity of SMSG was the fact that it was the first major project to be funded in the wake of the launch of *Sputnik,* the impetus that brought mathematical awareness to the public.

While SMSG was often criticized for trying to initiate a national curriculum, the writers stressed that they did not hope to push their materials on schools that did not want them. SMSG wanted to join in the efforts of other projects in providing an evolutionary program rather than a revolutionary program. However, they did believe that their curriculum was superior to the traditional one. They again stressed the hope for other publishers and groups to develop similar kinds of textbooks. SMSG also expressed the view that schools who "jumped on the bandwagon" and enforced a SMSG curriculum upon teachers unwilling to teach it would only cause disaster. "As a result of moving too fast, some schools attempting a wholesale transition to SMSG materials found themselves with a frustrated faculty, a confused student body, and a sadly disappointed administration"[79,p.140]. Critics of SMSG stated that the acronym stood for "Some Math, Some Garbage" [80].

SMSG designed textbooks so that a class could adopt a single textbook without having to adopt the entire series, and they were the first project to produce an entire high school textbook series. After this, commercial publishers became more important in textbook production, often following the lead set by SMSG. SMSG also produced films about mathematics for teachers to improve their own knowledge. Besides films, SMSG produced monographs on various topics and teacher guides to supplement their program. The large majority of SMSG's work was accomplished by 1961, and by 1965, most of its work was completed[81]. SMSG has been criticized for its speed in the production of textbooks that lacked a consistent underlying philosophy. Nevertheless, SMSG "was instrumental in continuing the revolution in mathematics education,"[82,p.76]. Many projects that developed after the beginning of SMSG followed similar ideas and formats as SMSG. Therefore, SMSG is often referred to as the leader and model of "new mathematics" projects—particularly for high school curricula as their elementary programs were not nearly as effective as their high school programs. Nevertheless, they were the first group to offer a complete and comprehensive "new math" program for grades K–12. SMSG also translated many of their textbooks into Spanish[83]. The popularity of SMSG can been seen in the fact that its materials were translated into fifteen languages[84].

SMSG published its first books for students of lower ability levels in 1962 for grades seven, eight, and nine. SMSG did eventually begin studying the effects of modern mathematics on disadvantaged children, but they were heavily criticized for doing it late in their program. SMSG did not investigate the "culturally deprived" child until 1965-1966. Much of what SMSG did at this point would not be remembered widely by the public because many had already formed their negative opinions of the "new math." Nevertheless, SMSG attempted to learn about problems disadvantaged children have such as lack of proper nutrition, supportive families, clothing, and other issues. SMSG concluded that no matter how great their materials or teachers were, teaching children affected in this way would not be as successful as they once thought. "SMSG erred by not including human concerns until late in the program, well after the textual materials and monographs were prepared"[85, p.191].

The Greater Cleveland Mathematics Project

The Greater Cleveland Mathematics Project (GCMP) began developing a modern mathematics curriculum in 1959 for students in public schools, grades K-12[86]. The project was directed by Dr. B.H. Gundlach of Bowling Green State and hoped to introduce new curricula gradually in all grades of elementary and secondary education after in-service training for teachers. This project dwelled on children understanding mathematics by using both logical structure and discovery learning. Guidelines for the program included: 1) the basic program must be suitable to all students; 2) the program must have a continuous flow of concepts for kindergarten through twelfth grade; 3) the program must be available to all students in kindergarten through twelfth grade; 4) the discovery method of teaching should be used as much as possible; and 5) the program must be mathematically correct and educationally valid. The GCMP stressed the logic and structure of mathematics to all students, and one study indicated that students using the GCMP curriculum performed at a higher level than students not using the GCMP curriculum. This program developed special learning programs for low-ability students. The GCMP stressed the use of their developed materials as well as the importance of teacher education in the new materials.

The GCMP never achieved its goal of effective curriculum change in K –12 grades, but it did become the most popular and well developed project concerning elementary "new mathematics" programs. GCMP developed materials for grades 1- 3 before any other "new math" elementary materials were published. GCMP then analyzed SMSG materials for grades 4-6 before developing their own materials for these grade levels. Testing of the materials was completed in local Cleveland suburban and parochial schools[87]. A major difference in this project was that funding did not come from the government, but instead was made possible by local foundations and businesses. In fact, the project was initiated by the Educational Research Council of Greater Cleveland. GCMP had developed materials for grades K-6 by 1965, but they failed to develop more materials because by 1970, many other "new mathematics" programs had been developed for older grades. Furthermore, the city's funding began shifting to other projects in the area.

The GCMP project is considered to be most in line with the UICSM project as both dwelled on discovery learning and teacher retraining programs. Discovery learning was regarded as the most appropriate way to teach this mathematics curriculum. The GCMP held retraining programs for teachers who were provided lectures by mathematicians and mathematics educators. These in-service trainings were important to the project, and the lectures were held in various parts of Cleveland to make teacher participation easier. The training also included daily television lectures. The teachers were given time off from work and credit towards advancement on the salary scale for attending the training programs. Besides developing curricular materials, GCMP also published books designed specifically for teachers to teach teachers the mathematics they would need in order to teach their own classes. GCMP also provided many materials to help teachers who were unable to attend training sessions. Overall, GCMP can be accredited with the most effort in educating teachers about "new mathematics".

While this project did influence teacher education with production of materials, information concerning this project is sparse. The lack of original material that can be found may be perhaps attributed to the fact that this project was funded by businesses in the area rather than originally based at a university—materials possibly could have been discarded instead of placed in archives.

The Syracuse University – Webster College Madison Project

The Madison Project of Syracuse University and Webster College (often referred to as the Madison Project) also sought to develop a mathematics curriculum, but developed it in stages[88]. The goal of the project was to create mathematical experiences for children that would be "better" than previous experiences, to make this known to the general public, and to understand what makes a more creative curriculum like this effective[89]. The Madison Project received its name from the Madison School in Syracuse, New York, where the initial work began[90]. The project was directed by Robert B. Davis and began in 1957 with focus on elementary mathematics. The project moved to Webster College in St. Louis, Missouri, in 1961 to focus on mathematics curriculum and teacher preparation. Various newspaper articles insinuated that grants had been obtained by Webster University to allow Davis to create materials and an elementary mathematics specialist master's program[91]. While the majority of the project's work was completed in classrooms with average or above average ability students, the first experimental year focused on low ability students, and the project had remarkable results[92]. Therefore, while many of the "new math" projects were criticized for not testing lower ability students, the Madison Project showed that its materials worked well with students of all abilities. While working with high ability students (which the Madison Project described as the upper 1/3 of the school population), the Madison Project proved that much of the mathematics learned in high school and college could be learned by bright elementary students.

The format of the project was as follows: grades K–3 would mainly learn the traditional mathematics curriculum; grades 4–7 would also learn the traditional mathematics curriculum with the addition of the Madison Project's materials on a one-lesson per week basis; and further grades would incorporate new materials at a steady rate. Therefore, the initial work was concentrated at the 4–9 grade level. Davis encouraged using the materials approximately one hour per week and stated that some materials could be used as early as grade 2 and as late as high school[93]. However, the materials were designed for grade school use because the leaders of the project wanted to utilize the natural creativity and curiosity of young children.

Some of the objectives of the Madison Project were: 1) the child should develop his ability to discover patterns in abstract situations; 2) the child should develop independent exploratory behavior; 3) children should begin to develop mental symbols for what they have explored; 4) children should learn and understand fundamental mathematical ideas; 5) the child should realize that he really can discover mathematics; 6) the child should develop critical thinking skills; 7) the child should regard mathematics as exciting, challenging, and worthwhile; and 8) children should begin to understand the importance of mathematics in history and their culture[94].

The purpose of the project was to improve the teaching of mathematics and investigate how and why students learn mathematics. Therefore, the emphasis was not on facts but on process of learning. Importance was placed on the development of materials and instruction methods for teaching by determining the best possible instruction for children of various ages and ability levels. The characteristics of the Madison Project were: 1) students learned from experiences in analyzing problems; 2) drill was eliminated and creative interest was encouraged; and 3) students were given more individual attention[95]. The Madison Project hoped to do this by concentrating on supplementary materials and teacher training rather than producing textbooks for all grade levels. Therefore, the Madison Project materials were to be incorporated into already existing "new math" programs[96]. Although most of the content was focused around only mathematics, some material was established to create an integrated math and science curriculum.

The project tested its materials in classrooms for many years, and research like this was only seen in one other "new math" project, the UICSM project[97]. This project strived to bring the "spirit" of "new math" to children by providing students with "real" experiences that excited them. In order to do this, the project had five objectives: 1) to provide reasons for mathematical manipulation; 2) to present children an opportunity to develop their mathematical thinking; 3) to create classrooms where children could actively participate in mathematics; 4) to teach more advanced mathematics to younger children than usual; and 5) to help teachers and schools in terms of training and time management.

The Madison Project also encouraged discovery learning and discouraged rote drill. Davis was credited for being an effective teacher using guided discovery in the classroom. This "discovery method" was described as: the teacher asking a question; students answering whether the concept was true or false; on that answer new questions were built, and this continued with very little explaining by the teacher; students were allowed to perceive the pattern that was obvious to them; and basic principles were established[98]. Thus, a class activity was led by the teacher but not dominated by the teacher[99]. The teacher should have developed a lesson plan, but it must not have been rigid. The teacher should have provided relatively informal situations and used informal language, and the teacher should not have shown the children how to attack the problem. Students should have been allowed to creatively explore mathematics[100]. Furthermore, activities should have been developed appropriate to the age of the child. For example, more abstraction should have been seen in older elementary grades than younger ones.

A major commitment of the Madison Project was to investigate Bruner's philosophy of discovery teaching as the authors felt improved instruction was as necessary as improved content. The project felt that the ability to discover patterns in abstract material was one of the most essential mathematics skills—if not the most essential skill[101]. Davis also encouraged the use of manipulatives in discovery teaching. By providing a positive learning environment depicting the "spirit" of mathematics, students wanted to participate and learn. In order to achieve such a classroom, Davis explained that children should have an active role by "inventing" or "discovering" techniques to solve problems in an informal setting[102]. Furthermore, experience with basic concepts should precede any formal instruction so that students began to realize their own ability to discover mathematics.

The Madison Project explained that modern mathematics could not be achieved by adopting new textbooks[103]. Instead, many other issues in education had to be addressed such as educational teaching philosophies, further teacher education, and the introduction of laboratory experiences in mathematics with less reliance on textbooks. Davis also saw the need for continual collaboration among university mathematicians and school teachers. The Madison Project stressed three working hypotheses: 1) experience with basic concepts should precede formal instruction; 2) discovering patterns in abstract materials should be the most essential mathematical skill; and 3) content and method are equally important and inseparable[104]. In general, parents, students, and communities were pleased and enthusiastic about the Madison Project.

While the materials and pedagogy developed and used by Davis were considered to be extremely useful in teaching children how to understand mathematics, Davis became bothered when seeing educated teachers unable to teach the Madison Project materials in an effective way using the discovery method. Therefore, in the early 1960s, Davis began a Master of Arts Teaching Program at Webster College specifically for elementary teachers who wanted to become mathematics specialists. This course involved the materials from the Madison Project and required teachers to take thirty credit hours related to mathematics and teaching[105]. The Madison Project also offered many in-service courses and training in its materials. The project even had a service whereby teachers could submit tape recordings of their lessons for critical evaluation[106]. Davis stated that only above average teachers could adequately challenge high ability students; therefore, the Madison Project began to focus more on teacher development[107].

This project saw the need to educate teachers by using films rather than texts. These films were designed to be viewed by teachers, parents, mathematicians, administrators, and psychologists[108]. They were to prove to others that students could learn higher level mathematics than we had been teaching. The films also showed how children "discovered" ideas themselves. Because these films were not "staged," they were effective by showing an audience an actual school experience where children were having fun discovering mathematics. Furthermore, if teachers or others were interested in following the students' progress through mathematics, they could watch many of the films that depicted the same students for as long as five consecutive years[109]. Therefore, in order to improve instruction, the project felt films could effectively show teachers the discovery method as well as an effective class dominated by a "new math" project.

The Madison Project also realized that in order for a school to successfully adopt any "new math" curriculum or any new curriculum for that matter, the following issues had to be addressed: there must be general parental and teacher agreement as to the goals of a new curriculum; the teachers must be creative, well-educated, and flexible; the school administration must be supportive and flexible; and the classroom and school must have a non-authoritarian atmosphere. The project felt that classes that were non-authoritarian produced more creative students who felt free to explore without fear of disappointment. Furthermore, these communities must have respect for children, mutual respect for parents, teachers, and children involved in the new projects, a general atmosphere of creativity, and general satisfaction of children's needs.

Concerning content, the Madison Project believed in the spiral approach to mathematics—a subject is not explored too much within a single lesson, but recurs periodically in various ways until it becomes familiar. The project hoped to build on topics in arithmetic; provide a basic foundation for integrating arithmetic, algebra, and geometry; provide a foundation to relate mathematics to science; and provide a program for the academically talented to move ahead more quickly in mathematical content. In testing the Madison Project as with many of the other "new math" projects, students performed well on both the traditional content and at a higher level of more abstract or "new" content[110].

The authors of the project felt that content and methods were equally important, and the most important issue was to assess what had been learned by the children. This included the attitudes that were learned as well as the content that was learned. While Davis did create many materials dwelling on content, the Madison Project offered much more to the "new math" movement in terms of teaching methodology and pedagogy. The largest difference between this "new math" project and others was that the Madison Project focused on improving actual classroom experiences rather than preparing textbooks, and encouraged teachers to establish a mathematics laboratory approach rather than a textbook-dominated approach.

"If we can educate children so that they will like mathematics, feel at ease with it, and approach it with powerful originality, we will have accomplished our objectives"[111, p.4].

Comprehensive School Mathematics Project

The Comprehensive School Mathematics Project (CSMP) was the last major project of the "new math," and while its roots began at the end of the "new math" movement, the final direction was very different than those projects dominant in the late 1950s and early 1960s. CSMP was mostly housed at Southern Illinois University in Carbondale, Illinois. It was designed to address specific issues stated in the Cambridge Conference report, *Goals for School Mathematics*, published in November 1963[112]. Furthermore, CSMP hoped to establish their project as a research based one—an issue that other projects had felt needed attention. CSMP started at Nova High School in Fort Lauderdale, Florida in 1963 under the direction of Burt Kaufman and moved to Carbondale in August 1966 under the sponsorship of the Central Midwestern Regional Educational Laboratory (CEMREL) which was established under the Elementary and Secondary Education Act of 1965[113].

The goals of CSMP were to modernize both the content in mathematics and the teaching methods of mathematics[114]. CSMP hoped to use some of the ideas from the earlier "new math" materials, and incorporate them into this new project[115]. CSMP desired to integrate mathematics topics by not teaching in distinct topics such as algebra, geometry, and so forth. This project stressed logic, mathematical structure, and precision in language[116]. CSMP also promoted discovery learning and problem solving with less rote memorization than the traditional curriculum, and the success of the program was hoped to be achieved through adequate teacher training. They realized that teachers would need a stronger background in mathematics as well as the ability to work with students individually and in small groups as a tutor, evaluator, and resource. However, in reality, the CSMP put more emphasis on mathematics content than pedagogy. CSMP did train coordinators and teachers in workshops during the summer as well as provided guidance to those teachers using the materials[117]. They were aware of teacher training limitations and discouraged the use of its materials by teachers who had not been trained specifically by CSMP.

The main goal of CSMP was, "to develop mathematics curricula for students from ages 5 through 18 which provide for each student a program sound in content, enjoyable, appropriate to his needs and abilities, and presented so as to maximize his success in the realization of his potential for learning and using mathematics"[118,p.12]. CSMP believed that mathematics should be taught as a unified whole; therefore, they taught their topics in a spiral curriculum—one where topics are reintroduced and taught to a higher level periodically[119]. The materials were designed for individualized learning (learning based on the needs of the student) either for a whole class, small group of students, or individual student[120]. Besides the general project materials, CSMP also developed supplementary materials for slower learners, but developed an entire course called Elements of Mathematics designed for gifted students (described by CSMP as the upper 10-15% of the student population). CSMP realized that not all students would begin their first year of college with calculus. They saw the need for some students to develop more computer skills, and others in fields such as business and social science to focus on topics such as statistics. CSMP stated that students needed Advanced Placement exams in other areas of mathematics besides calculus such as statistics and linear algebra.

Some of the basic objectives of CSMP were: 1) to analyze the structure of mathematics in order to outline content and abstraction levels of elementary mathematics; 2) to develop individualized mathematics programs; 3) to develop activity packages suitable to content; 4) to develop in-service and pre-service teaching programs in connection with this individualized curriculum; 5) to develop methods for evaluating the program; and 6) to develop a system to manage this individualized curriculum in a school[121]. Studies completed pertaining to the level of achievement of students using the CSMP materials showed that they were at or above the level of non-CSMP students with regards to traditional material, and their reasoning ability was much better than non-CSMP students.

While the CSMP did not associate itself with the "new math" movement (most likely because of the backlash that had begun), many of its ideas were very similar to the "new math" ideas. These included encouraging understanding of mathematics and creativity of students. Despite its widespread use, because CSMP began in the mid-late 1960s, it never became dominant as a "new math" program because of the strong hold SMSG and other existing "new math" programs had on the public.

A major difference between CSMP and other projects established during the "new math" era is that CSMP continued to develop materials through the 1990s. CSMP changed its name various times due to alterations in funding sources, and often its focus and curricular materials changed throughout the decades so much that most do not associate it with the start of the "new math." CSMP became very popular in the 1970s as an alternative to a "back to basics" curriculum that a few schools tried to avoid. CSMP developed a strong history over several decades; therefore, the evolution and changes based on the original project is fairly easy to follow. This is not the case for many "new math" projects. Another issue is that many of the projects of the 1950s and 1960s ended in their present form when their leaders left or changed focus. However, this did not mean that the projects ended—they often evolved into other projects. Furthermore, many of the later changes were not linked to the original projects making it difficult to understand the evolution and changes of the projects over time.

Minor Projects

The minor projects described next are labeled as minor because they did not necessarily have the same influence as the major projects. However, one of the reasons for not having as much influence is because the projects continually evolved as different projects often under different leaders. Without the consistent push of one curriculum or philosophy, these minor projects, while important in contributing to the "new math" movement, were not considered as the driving forces of change in schools.

Developmental Project in Secondary Mathematics at Southern Illinois University

The Developmental Project in Secondary Mathematics at Southern Illinois University in Carbondale, Illinois, under the direction of Morton Kenner, was funded to specifically implement the CEEB's recommendations[122]. This program began developing materials for high school students, and planned to attack the development of junior high school materials once the high school materials were established. The program emphasized the precision of language and mathematical structure. This project developed some high school mathematics textbooks that were tested in the University High School including topics such as algebra, sets, number theory, and statistics. The texts were criticized for introducing topics with no explanation and for incorporating little use of discovery teaching. The texts did not clearly indicate the audience for which they were developed, and they did not incorporate much teacher or supplementary guidance. This project's materials did not reach wide circulation, and consequently, did not have much influence in the "new math" movement. However, its director, Dr. Kenner, began a "new math" program in a Mathematics Center in Nairobi, Kenya, Africa, in the early 1960s under the support of Educational Services Incorporated[123].

Boston College Mathematics Institute

The Boston College Mathematics Institute (BCMI) became a formal organization in 1957 under the direction of Rev. Stanley J. Bezuszka with funding from the National Science Foundation[124]. This project began with the intention of developing texts for grades 8 –12 with an historical development and point of view. The project eventually hoped to develop a curriculum where students had completed the traditional algebra content by the end of eighth grade and the traditional high school curriculum by the end of eleventh grade— leaving twelfth grade available for the teaching of calculus and computer based applications.

The project eventually produced junior high mathematics texts and a "modern math" text for teachers. The institute attempted this by promoting the re-education of high school teachers in modern mathematics[125]. The institute developed a text titled, *Sets, Operations, and Patterns, a Course in Basic Mathematics*. The text encouraged discovery teaching as the pedagogical method for the concepts included such as sets, number, inequalities, factoring, and basic operations in algebra. They also developed computer program courses for high schools. The institute shifted its focus from production of textbooks to more teacher training and collaboration with commercial textbook producers. In fact, "new math" was accepted in teacher training institutions because of the work at the BCMI.

Other projects that dealt specifically with improving teacher training and providing in-service training as well as opportunities for critical discussion and analysis of "new mathematics" programs included the Illinois Curriculum Program Thinking in the Language of Mathematics and the Indiana School and College Committee on Mathematics[126].

The School Mathematics Project

The School Mathematics Project (SMP) in England focused on developing a modern mathematics program for students aged five to thirteen[127]. This project promoted a more open learning environment in the teaching of mathematics and was developed solely by teachers with little input from university mathematicians. Modern mathematics topics followed the same lines as those seen in the United States, but national funding was limited to the projects in England. SMP was founded in the early 1960s to produce new texts and examinations which the authors felt would encourage new approaches to teaching[128]. In general, their work was readily accepted by teachers because it was a movement driven by teachers' desires to improve mathematics education in the United Kingdom. SMP materials were used in over 50% of English schools[129].

SMP developed a modern syllabus for the national curriculum in the United Kingdom; however, the authors planned a continuously changing syllabus to meet the varying needs of the students[130]. Furthermore, SMP stressed that reform needed to begin in the primary grades rather than the upper levels in schools. The major aims of SMP were to develop a mathematics curriculum that was more enjoyable for students as well as incorporated the nature of modern mathematics. The SMP further stated that the increase in technology (both automation and computers) made revising school mathematics necessary. Furthermore, SMP hoped to develop a curriculum that encouraged more students to pursue further education in mathematics[131]. The curriculum seemed to follow those already developed in the United States by incorporating new topics such as sets and matrices with the discovery method as SMP felt there were "serious shortcomings in traditional school mathematics syllabuses"[132,p.i.]. SMP also promoted teacher institutes where their materials could be taught to in-service teachers.

Nuffield Mathematics Project

The Nuffield Project was centered in England and developed a program based on children doing mathematics and actively learning[133]. The Nuffield mathematics project began in 1964 and was supported by the Nuffield foundation. The project hoped to devise a contemporary mathematics curriculum for students aged 5–13. This project was a non-textbook based project similar to that of the Madison Project[134]. The Nuffield Project also saw the need for modern mathematics in our society in order to meet the needs of today and the future because of computers and automation[135]. This project allowed elementary aged children to investigate problems working together in teams using everyday ideas and topics that were interesting. This project also stressed the use of manipulatives in learning. The goal was again understanding of mathematics, and the Nuffield Project supported the idea of the spiral curriculum.

The Nuffield Project did prepare books for teachers, but no textbooks were used in classroom instruction. In order to train teachers, the Nuffield Project developed "teacher study and preparation centers" around England where teachers could collaborate in a relaxed atmosphere. Sample curriculum materials and teacher reference books were provided in centers so that teachers could develop lesson plans based on these resources as opposed to following a textbook.

The Ontario Mathematics Commission

The Ontario Mathematics Commission formed in 1959 in Ontario, Canada, and began developing a secondary mathematics curriculum under the direction of Frank C. Asbury[136]. Its goals included: 1) encouraging experimental teaching materials and the testing of those materials; 2) cooperating with universities to provide teacher education courses of a higher standard than present day; and 3) developing the best curriculum practices and making the information widely available. In reaching the goals, the Ontario Mathematics Commission stated that the curriculum must be continuous and must be implemented with appropriate materials and by appropriately trained teachers with an interest in developing students in mathematics at all levels. More importantly, the commission stressed the idea that the development of a curriculum would be a long-term process involving both high school teachers and university professors. The commission also stated that the key to successful curriculum reform was the preparation of teachers.

The Minnesota Mathematics and Science Teaching Project

The Minnesota School Mathematics and Science Teaching Project (MINNEMAST) formed in 1961, and was first headed by Paul Rosenbloom and then by J.H. Werntz[137]. Paul Rosenbloom was director of its forerunners, the Minnesota School Mathematics Center which supported joint work by teachers and college teachers and the Minnesota National Laboratory for the Improvement of Secondary

School Mathematics which tested existing "new math" programs, particularly SMSG materials, in various schools. The goal of the project was to develop a coordinated mathematics and science curriculum for grades K-9 and science and mathematics methods courses for pre-service teachers.

The curriculum addressed elementary school students and emphasized the following: 1) to understand how a scientist acquires knowledge; 2) to observe, classify, generalize, and predict; 3) to recognize problems and discover their answers; and 4) to build a structure of scientific knowledge. MINNEMAST's main goal was to develop an integrated program of science and mathematics for grades K–9, and this project was also funded by the National Science Foundation. This project also stressed discovery learning where the child could discover, observe, and explore mathematical patterns. The science team and the mathematics team developed materials separately at first. This group also produced teacher training materials.

The science and mathematics materials did not come together until the late 1960s, and funding was suspended by 1972. Therefore, the group only produced materials for grades K–3, and was not popular because many other projects had already developed materials for grades K–6. However, this project is unique in that mathematics was taught around the learning of natural science. Furthermore, the materials developed for ages K–3 were extensive, and children were given books to take home so that parents could be more involved with this new project.

Ball State Teachers College Experimental Project

The Ball State Teachers College Experimental project was led by Charles F. Brumfiel[138]. This was one of the first projects to develop after UICSM. Testing of the Ball State project's geometry materials began in 1955, testing of algebraic materials began in 1957, and testing of eighth grade materials began in 1958[139]. The original goal was to produce texts for grades 7–12 that were abstract but could be taught well. The group produced a general mathematics text, an algebra I text, and a geometry text. All emphasized logic and the deductive structure of mathematics with little emphasis on drill. The original materials were tested in the Ball State Laboratory School, Burris, and other schools in the area. The students who were tested were described to have average ability. Teachers had little formal training in the project materials but met periodically to discuss pedagogical techniques. The more capable students and teachers enjoyed the project from early analysis. Weaker students performed equally as well as in traditional programs, and average students performed quite well—particularly those who were highly motivated[140].

Even though the materials from the Ball State project never dominated the textbook market, the project did prove that a program containing mainly abstract mathematics could be taught successfully to secondary level students. Furthermore, the Ball State project became less popularly known as the "Ball State Project" because Addison Wesley, the publishing company, took over the writing efforts and began publishing their own materials based on those developed at Ball State[141]. Addison Wesley produced materials covering grades K–12 based on work initiated at Ball State by Brumfiel, Eicholz, and Shanks.

The Suppes Project

Another project was an elementary mathematics project centered at Stanford University under the direction of Patrick Suppes[142]. The project did not have an official name and became known as the Suppes project. This project focused on experiments in teaching various new mathematics topics to young children. The program specifically taught set theory in connection with arithmetic in order to make arithmetic more meaningful and logical. Another aspect of the program based at Stanford was the teaching of geometry. Another project based at Stanford taught mathematical logic to gifted students. The Suppes project focused on first through third grade in California—much of the work was completed at the Stanford Elementary School. Early results of teaching this material to young children showed that the students' vocabulary and reading skills improved, and they learned the material quickly. Suppes produced textbooks for grades K- 6 by 1966, and all material was based on what had been taught in the many programs led by Suppes at Stanford. Therefore, besides the elementary material produced by SMSG and GCMP, the Suppes textbooks were some of the few designed for elementary "new math" programs.

Besides the minor projects described previously, many universities, states, local school districts, and even local schools developed their own mathematics curriculums, curriculum philosophies and statements, and curriculum materials. While none became as popular as previously mentioned projects, the influence of these many various small projects on the "new math" movement helped build the major reform movement during the 1950s and 1960s. Examples of a few of these are explained below.

State Mathematics Syllabus Committees
          The New York State Mathematics Syllabus Committee under the direction of F. Eugene Seymour was established in 1942. It was a state committee charged with establishing a mathematics program to meet the needs of all high school students[143]. Its suggestions followed the recommendations of the Commission of the CEEB. In fact, many states formed their own committees to work specifically on a mathematics curriculum for their own state. Examples of these groups are the Oklahoma State Committee on the Improvement of Mathematics Instruction and the Commission to Study the Mathematics Curriculum in Texas Elementary and Secondary Schools. Another state that began to be involved in the "new math" was California. The California State Curriculum Commission selected appropriate textbooks for the state. This commission was genuinely interested in developing criteria for the improvement of mathematics education throughout the state. They stated that issues such as researching experimental programs, improving instruction and curriculum in mathematics, improvement in teacher education, and improvement in teaching methods needed attention.

Algebra for Grade Five
          Algebra for Grade Five was a program developed by Dr. W.W. Sawyer at Wesleyan University in Middletown, Connecticut. This program developed an algebra unit for fifth graders who were performing in the top 33% of their class. The students studied algebra one day a week, and the program was originally intended to continue for grades 6, 7, and 8. The leaders of this program wanted to prove that more capable students in the United States could begin studying algebra at the same age as the more capable students in Europe. However, this project stressed that algebra at this age can only be effective if taught in a meaningful way through discovery and invention. The project hoped to help those students who became bored with arithmetic. Early results showed that students found this new work on algebra interesting and comprehended it easily as well as improving their arithmetic skills.

Mathematics Program at Phillips Exeter Academy

Another program was the Mathematics Program at Phillips Exeter Academy in Exeter, New Hampshire, where the school mathematics staff developed a mathematics program over a twenty-five year period. This project also developed a curriculum along the lines of the CEEB's commission. The mathematics staff at this school prepared a series of mathematics textbooks for grades ten through twelve that became available for purchase by other institutions[144].

Illinois Curriculum Program Study Group on Mathematics

The Illinois Curriculum Program Study Group on Mathematics was a group interested in the overall improvement of mathematics education K–16[145]. This group called for elementary teachers to teach for understanding rather than skills only. They also felt that students at the junior high level were re-learning elementary mathematics unnecessarily, and that heterogeneous grouping was limiting the academically talented. Concerning high school mathematics, this group felt that traditional mathematics did not adequately prepare students for college mathematics, and that gifted students were not being challenged enough. In general, this group believed teachers were inadequately prepared to teach mathematics and that schools and teachers had not kept up with the changes. This group suggested that: colleges review preparation requirements for all mathematics teachers; mathematics classes be ability grouped; and at least one qualified teacher and adequate sets of materials be provided in each school to establish a high mathematical ability class. Concerning college, this organization stated that students should begin instruction with calculus and analytical geometry and that modernization of the undergraduate program was necessary.

Besides mathematics projects, many science-based projects were being established at the same time. The first major project was the physics project, the Physical Science Study Committee (PSSC), discussed at the Cambridge Conference that started the formation of SMSG as explained previously. Chemistry and biology groups also started developing modern programs. Another project was the Elementary School Science Project sponsored by the University of California, Berkeley. The work of this project began in 1959 and focused on guiding children to form fundamental scientific concepts that met their needs while presenting science from a professional viewpoint.

While a comprehensive list of major and minor projects related to the "new mathematics" movement has been provided, it would be impossible to discuss all new programs developed in the 1950s and 1960s. The ones listed are normally more well known and provided new information and materials longer; however, there are many projects that were developed on a smaller scale. Many universities and city school districts designed their own "new math" programs. Some of these only operated briefly or in one school or area. In fact, it was estimated that over six hundred programs could have fallen under the title "new math"[146].

Furthermore, many writers of textbooks and materials for one project were involved in other projects as well. For example, John Mayor who was the leader of the UMMaP project helped SMSG develop some of their materials. Dr. Eicholz who worked on the Ball State project was involved in the Greater Cleveland Mathematics project too[147]. Also, some leaders of UICSM acted as a consultant to GCMP. This contributed to many national projects and smaller projects having the same formats.

Curriculum Groups

Besides the many university, state, and local projects or commissions, various important curriculum groups and conferences affected the "new math" movement significantly. While curriculum committees existed throughout the country, the following national committees produced reports or statements affecting "new math" throughout the country.

Cambridge Conference on School Mathematics

Two conferences considered more radical were the Cambridge Conference on School Mathematics and the Secondary School Mathematics Curriculum Improvement Study[148]. The Cambridge Conference on School Mathematics (CCSM) recommended that the K-12 curriculum achieve a level equivalent to three years of undergraduate mathematics, and assumed that teachers should be able to adjust with relatively short training. This group believed that students at the end of high school should have studied two years of calculus, one semester of modern algebra, and probability theory.

This Cambridge Conference was the second of the same name. The first Cambridge Conference was considered one of the key elements in the formation of SMSG. This second Cambridge

Conference, called the Cambridge Conference on School Mathematics (CCSM), produced the report, *Goals for School Mathematics*, in 1963. The report, based on the conference attended by twenty-nine mathematicians from universities and industry, concentrated on improved mathematics content with little emphasis on pedagogy. One positive aspect was that prominent university researchers and mathematicians were interested in improving K-12 education. The members at the conference were not planning mathematics curriculum for the present but were instead planning for the future—particularly they were looking toward the year 2000. Specifically, they wanted to push down mathematics even more so that students entering college would have completed two years of calculus, one semester of probability and statistics, and one semester of linear algebra[149]. The CCSM believed that "pushed down" mathematics could be achieved by major reorganization of the elementary school curriculum.

The report from the second Cambridge Conference briefly addressed the issue that understanding of mathematics should be a goal rather than traditional drill[150]. The report stated that the discovery approach is invaluable in developing an understanding of mathematics, but the major problem with it was the time needed to use this technique. The report clarified that aided discovery (guided discovery) is a much more effective means of teaching mathematics for understanding in the time allotted in school. However, the report also stressed that direct instruction has value as well and should be incorporated when necessary. The committee hoped to achieve this by offering mathematics in a "spiral" approach where topics are reintroduced each year with more sophistication—they felt this would provide a clear picture of mathematics as a continuously growing field. While some of the ideas of challenging the mathematically talented as expressed in the report are valid, many educators felt that the report offered high expectations without the consideration of how children learned mathematics—particularly in light of the new educational research in this area. The report also ignored the problem of teacher training.

The third Cambridge Conference was held in 1966, and the subject of this conference was teacher training in order to be able to implement the goals that were established at the second Cambridge Conference. The name of this conference was the Cambridge Conference on Teacher Training[151]. Some influential mathematics leaders designed the content of the conference: Edward Begle

(SMSG), Robert Davis (Madison Project), and David Page (UICSM)[152]. Several American school teachers and British mathematicians attended the conference as well as university professors.

Much of the work of this conference dwelled on improving elementary mathematics and methods courses for pre-service elementary teachers. The members of the conference assumed that elementary pre-service teachers would enter the university with the equivalent of a tenth grade SMSG background. They recommended that the pre-service course for elementary teachers include mathematics training in arithmetic and number theory, logic, probability and statistics, geometry including transformations, conic sections, algebra including linear equations, quadratic functions, matrices, and elementary trigonometry[153]. This conference recommended that elementary teachers be given more free periods to allow them to prepare adequately for mathematics classes and even take in-service courses. Many of the objectives for elementary teachers were considered too high a level and extreme for the education of all elementary teachers.

Secondary School Mathematics Curriculum Improvement Study

Howard F. Fehr began the Secondary School Mathematics Curriculum Improvement Study (SSMCIS) inspired by the Cambridge Conference that also affected the CSMP project[154]. The SSMCIS attempted to integrate the entire mathematics curriculum and eliminate the division of separate subjects such as geometry, algebra, trigonometry, etc. The Secondary School Mathematics Improvement Study wanted to unify several branches of mathematics in order to bring college level mathematics into secondary schools.

The SSMCIS started its work in 1965 with a focus on the academically gifted in mathematics (namely the top 15 to 20 per cent)[155]. This group hoped to bring the equivalent of one year of university mathematics into schools. This group produced syllabi that were tested by teachers in schools in the mid 1960s. The SSMCIS was following European reform movements and models, and called for a very advanced mathematics curriculum to be completed by the end of high school. While this program did achieve some success, it was never widely adopted.

National Association of Secondary School Principals

A report of the National Association of Secondary School Principals, the Place of Science and Mathematics in the Comprehensive Secondary School Program, was printed in 1958 and recommended a curriculum sequence in science and mathematics for junior high schools and high schools[156]. Concerning the junior high school, the report stated that science and mathematics needed to be brought up to date, and all pupils should have some mathematics through grade 9 including not only arithmetic, but also some algebra and geometry. Concerning the secondary school, this report emphasized the need for all children to have a sound basic education related to their culture. This National Association of Secondary School Principal's publication paralleled many "new math" philosophies by recommending that students have the option of taking algebra, geometry, intermediate algebra, trigonometry and elementary analysis in three years leaving the fourth year option available for analytic geometry and calculus.

National Council of Teachers of Mathematics

The National Council of Teachers of Mathematics (NCTM) also supported the "new mathematics" reform, particularly through the *Mathematics Teacher* journal[157]. The popularity of NCTM increased dramatically during the "new math" era. NCTM established its own curriculum committee called the *Secondary School Curriculum Committee* (SSCC), and its report in 1958 and 1959 paralleled the CEEB's report about mathematical reform. This sixteen-member committee included members from both universities and industry, and the chairman was Frank B. Allen[158]. The purpose of the committee was to study mathematics curriculum and instruction in secondary schools and propose solutions for strengthening and improving the mathematics curriculum throughout the United States for grades 7-12. Some of the issues in this report included the place of mathematics in a changing society; the aims of mathematics education and the pedagogy of mathematics; the nature of mathematical thought for students; the content and organization of junior high mathematics; mathematics programs for students of all mathematical abilities; gifted mathematics programs; teaching aids; and the organization of mathematics programs. Many of the original "new math" projects were active in causing reform before NCTM published its report.

The report by the SSCC was very detailed—particularly describing the findings of its ten subcommittees. A brief summary of the title of the subcommittee and its findings is provided:

1) The place of mathematics in our changing society:
   Knowledge is increasing as most of what we know is relatively recent, and automation and the computing machine place new requirements on necessary mathematics knowledge.

2) Mathematics programs in Europe :
   The European system involves specializing and training earlier in a student's career, and teacher training is more selective in Europe.

3) The aims and pedagogy of mathematics:
   Classrooms should be more student-centered while indicating desired behaviors and learning activities. More emphasis must be placed on understanding and broad concepts, and less on drill and practice.

4) The nature of mathematical thought:
   More emphasis should be placed on the discovery method with a modern curriculum consisting of new points of view as well as new topics.

5) Junior high school mathematics:
   New programs provide much more emphasis on algebra and geometry as psychologists believe that students at the junior high level are capable of working at a much higher level than traditionally.

6) Geometry:
   Elements of geometry should be taught throughout junior high school and high school.

7) Different ability levels of students:
   The same mathematical topics should be taught to all students, but the depth and complexity should vary depending on the ability of the students; thus ability grouping is desired.

8) Aids in teaching:
   A teaching aid pamphlet was established by NCTM that was intended to be used as a guide for the use of teaching aids in the classroom.

9) Organization of mathematics:

A "new mathematics" curriculum must reflect topics available to all children and be uniform and flexible. Grade nine should offer algebra, and students in grade twelve should have some choice in advanced courses.

10) Administration of mathematics programs:

Ability grouping should be encouraged, and counselors should be available to advise proper placement of students. Competent teachers are required, and in-service education was encouraged.

NCTM attempted to analyze some of the "new math" projects on the basis of social application, placement and structure, vocabulary, methods, proof, and evaluation in order to credit and support much of the research towards improvement of mathematics in schools. In one study by NCTM, teachers and students discussed the new projects as an improvement on the traditional mathematics curriculum. Both students and teachers stated that they favored the "new math," but teachers did warn that much more preparation and motivation was necessary on their part to benefit the students. Furthermore, many teachers commented that while slower students were not negatively affected by the "new math," it was apparent that these slower students did not improve in mathematics[159]. The UMMaP course was commended for use with all ability groups, and special editions of SMSG materials were written for non-college bound students. Some educators stated that less-able students can do well in the "new math" programs but may need to learn the material at a much slower pace than normal. Thus, the issue of addressing "new math" for all ability levels was present from the first interviews with students and teachers.

In terms of implementing a program into a school or classroom for the first time, it was suggested that a mathematician or a teacher with a very solid mathematics background be available for support to teachers teaching "new math" materials for the first time[160]. Support needed to be in the form of adequate preparation time as well as instruction not only in content but also in teaching technique. The training of teachers in mathematics content and new approaches in teaching could be most challenging but was necessary for a new program to succeed. Districts had to be willing to pay for teachers to participate in summer institutes or other educational means. Informing parents of changes in mathematics education was also seen as imperative to a "new math" program's success in a school. Another suggestion on implementation of a program was that if a program was started in a particular grade, provisions should be made to ensure the new program continues through twelfth grade.

In order to successfully implement "new math" programs in California public schools, the California State Department of Education offered the following guidelines to administrators of any school interested in implementing a "new math" program: 1) familiarize yourself with literature in the field; 2) provide teachers with available literature and resources; 3) urge teachers to participate in national professional organizations; 4) provide resources such as mathematics supervisors or department heads; 5) provide adequate time for teachers and personnel in implementing the program; 6) encourage teachers to participate in summer workshops; 7) enroll in programs that offer in-service courses or a mathematics consultant; 8) provide good working conditions such as smaller classes and release time for teachers; 9) group classes according to ability; and 10) maintain good communication with parents and other teachers[161].

Similarities among the projects

In order to analyze the "new math" movement of the 1950s and 1960s, similarities among the many projects must be established that incorporate the entire "new math" reform movement. Furthermore, in order to analyze the influence of the "new math" movement with the current reform movement, general characteristics of the 1950s and 1960s projects must be explained. The following list shows the similarities among the projects:

- The programs attempted to unify themes of the new mathematics, and most included new topics such as set theory, graphical representation, statistical inference, probability, structure, measurement, number, operations, and logical deductions.
- In adding new topics, obsolete topics were deleted such as solving plane and spherical triangles by logarithms in trigonometry and a number of theorems in geometry. In other cases the older topics were revised and incorporated differently.
- More math was being taught in less time.
- Most programs dwelled on the logical structure of mathematics and teaching students concepts essential for truly

understanding mathematics rather than traditional rote or drill and practice.

- Early available data supported that students involved in the programs performed as well as students on traditional tests and they understood basic principles better.
- Most projects involved first rate scholars in an alliance between colleges and schools.
- The projects were national in scope and were supported financially on a large scale.
- All dwelled on what should be taught with a desire to bring the curriculum up to date—particularly in a more advanced and technical world.
- Most developed educational materials with explorations into new and different ways of teaching, many incorporating the discovery method—casting the teacher into a different role. Students often worked together in groups and with manipulatives in solving problems. Many projects published their own textbooks throughout the movement, while others began publishing their own but were eventually published by educational resource publishers.
- All programs eventually developed textbooks as curricular materials, and most sold mimeographed versions of the textbooks long before they were bound and circulated by various publishing companies.
- Most of the larger projects eventually developed films of teachers using their materials to facilitate the problem of teacher training.
- Almost all projects supported the need for teacher training programs and materials as well as extra time or a lighter load for teachers initiating a program as most teachers were not adequately prepared to teach a "new math" program.
- Most projects followed the guidelines established by the CEEB that stated the mathematical course program in high school should be algebra, geometry, advanced algebra and trigonometry, and a twelfth year course including probability and statistics, functional analysis, or analytical geometry with possibly some calculus.
- All were the first steps in a long-range effort to improve the mathematics standards in schools in the United States.

By 1963, NCTM had praised SMSG and UICSM for their work, and they stressed the new curricula as explained by the CEEB and NCTM in *The Mathematics Teacher*. Both SMSG and UICSM placed heavy emphasis on the structure of algebra, precise language, sets, and the difference between the number and numeral. The UICSM and Madison projects were the two projects that led to the production of effective discovery learning materials; however, their materials were never in widespread use. While all the "new mathematics" projects continued the push towards a revised mathematics curriculum across the United States, SMSG as well as later commercial publishers had the most affect because of the large number of schools exposed to their materials.

Effects and Aftermath of the "New Math"

Chapter Four begins with an explanation of the educational changes that occurred both for students and teachers as a result of the "new mathematics" movement. The chapter continues by discussing the disappearance of the "new mathematics" movement in the late 1960s and early 1970s and offering some suggestions for why the "new mathematics" movement lost favor with the public. A short analysis of mathematics reform between the end of the "new math" movement and the present day is provided.

Educational Changes in the 1960s/70s

Many educational changes took place in the curriculum at the elementary and secondary levels, as well as in teacher education as a result of the "new mathematics" projects. One of the major differences between the traditional curriculum and the "new mathematics" curriculum was the use of modern language. Also, the language of sets and abstraction became part of the "new mathematics". Both set theory and abstract algebra became known synonymously with "new mathematics," but neither was a large part of the curriculum.

K-12 Changes

The traditional curriculum for all K– 12 education prior to the "new math" movement was one of routine computation and drill and practice. The traditional curriculum in elementary education was based completely on arithmetic. Modern mathematics in elementary schools presented new topics and terms that had traditionally been left to older students. Such topics as binary numbers, set theory, Venn diagrams, clock arithmetic, and ancient numerals of Egypt were included[1]. The elementary mathematics revolution focused also on integration within the curriculum, an increase in abstract thought, and discovery learning. The largest problem with reform in elementary mathematics was that the increase of project materials and recommendations in a school did not necessarily mean that the classroom content and teaching were improved.

The traditional curriculum at middle school or junior high level was arithmetic sometimes with basic geometry and a small amount of algebra. In fact, many regarded the 7th and 8th grade mathematics curriculum as a "mathematical disaster area" because of the continual repetition of arithmetic[2]. The typical traditional curriculum in high school was algebra in 9th grade, plane geometry in 10th grade, and intermediate algebra in 11th grade accompanied sometimes by physics, trigonometry, or solid geometry. Twelfth grade mathematics varied greatly with many courses including a review of all mathematics and some courses dwelling most on advanced algebra and sometimes solid geometry.

"New mathematics" hoped to move at least the first two years of college mathematics into high schools after traditional high school topics had been spiraled down into elementary and middle school[3]. This would allow a curriculum including more advanced courses like analytic geometry, calculus, modern algebra, linear algebra and matrices, and probability and statistics. In order to do this, many schools combined solid and plane geometry into one course and advanced algebra and trigonometry into one course. Computer programming also became popular as a course. This new curriculum stressed the need for improved problem solving skills. As part of promoting the use of critical thinking and problem solving skills, the "new mathematics" textbooks tried to incorporate logic and deductive reasoning throughout.

Teacher Standards Changes in the 1960s/70s

The importance of increasing educational standards for teachers was supported by mathematicians and mathematics educators by the 1960s. Four main reasons for improving teacher certification were: 1) the many "new math" projects changed the standard mathematics curriculum; 2) the promotion of "discovery teaching" and stress on creativity among students in mathematics; 3) more emphasis on the college-bound student as more students were attending college; and 4) the introduction of newer mathematics and higher level mathematics such as probability and statistics, linear algebra, calculus, etc. [4]

Content

Teachers frantically went to conferences and workshops to try and keep up with the changing curriculum. Some of the professional journals even published articles that explained some of the "new mathematics" topics as a means of educating teacher. Most of the project leaders wanted to train teachers in the use of new materials, and workshops became prevalent. UICSM trained teachers more than any other project, but almost all project authors desired, but did not require, some teacher training in the use of their materials. Most teacher training in-service programs were sponsored by the National Science Foundation. While many trained in summer workshops where their expenses were paid and they received a small stipend, some teachers completed year long courses at universities[5]. Some even began applying this extra training towards a master's degree. Another popular teacher training and in-service training program was established by Robert B. Davis at Syracuse University, and it was called the Syracuse Plan or the Sloan Experiment (as it was financed by the Alfred P. Sloan Foundation). In this plan, selected secondary school mathematics teachers taught approximately half of their usual teaching load and studied the other half of the time at a local university; however, they were given their full-time salary.

In 1954, the University of Washington held one of the first summer institutes for high school teachers in mathematics. By 1968, there were 458 institutes in summer[6]. Summer institutes were popular for both high school and elementary teachers, but far more high school teachers were trained in these programs than elementary teachers. Another major problem with the re-education of teachers was the large numbers of teachers. In 1965, there were an estimated 135,000 high school mathematics teachers and 1,100,000 elementary school teachers expected to teach mathematics[7]. Although the NSF institutes could reach a majority of secondary school teachers, it would be impossible to reach even a small portion of elementary teachers. However, the NSF surprisingly offered many more secondary institutes than elementary institutes despite the many more elementary teachers who would need further education. By 1965, it was estimated that half of all high school teachers had attended one or more National Science Foundation institutes[8]. The other half had not ever applied to attend a

NSF summer institute. It is estimated that between the years of 1961 and 1966, only approximately one per cent of all elementary teachers had been trained in a summer session. However, many of the institutes did not offer help in pedagogy or even content that the teacher would teach—instead the content dwelled more on developing even higher levels of mathematics (beyond what would be necessary to teach the new curriculum). Furthermore, as the Vietnam war escalated, government money that had been going to the in-service education of teachers began to be used for the war effort[9]. Consequently, the number of institutes greatly decreased.

Most "new mathematics" started in high school and filtered down to elementary school, and the lack of adequate teacher training and preparation, in particular at the elementary school level, remains the largest criticism of the "new mathematics" movement. The education of elementary school teachers was so lacking in mathematics that making at least two years of high school mathematics a college entrance requirement for prospective elementary teachers was recommended. At the start of the "new mathematics" movement, admission to teacher training programs was lenient—only three-fourths of all colleges required at least one year of high school mathematics. In 1950, thirty-five states required no college courses in mathematics for elementary certification, and as late as 1960, twenty-nine states still did not require college mathematics. Also by 1960, many institutions would admit students to the elementary education program who did not have any high school mathematics[10]. Luckily, some state universities required mathematics for graduation.

It was further recommended that secondary mathematics teachers be required to complete twenty-four hours or six courses of mathematics beyond calculus, and have one minor area of concentration in a mathematics related field as well as education classes. Recommendations for courses beyond calculus that would be appropriate for secondary teachers were: history of mathematics, advanced calculus, probability and statistics, modern algebra, differential equations, logic, and geometry[11]. State requirements were also minimal for secondary education certification. Before 1960, only fifteen hours of college mathematics was required to be allowed to teach mathematics at the high school level, and teachers who taught mathematics only were required to have twenty-five hours of mathematics in college[12]. This number may seem adequate, but before 1960, most students entered college with a lower level of mathematics than today, and the majority of hours required only included trigonometric function and calculus—thus high school teachers were graduating with little mathematics knowledge beyond calculus.

To help develop a higher standard for mathematics teachers in elementary and high school, the *Committee on the Undergraduate Program in Mathematics* (CUPM) was developed by the Mathematics Association of America in 1960 and was formed from the original Committee on the Undergraduate Mathematical Program. The original Committee on the Undergraduate Program (CUP) was established in 1953 to act as a connection between research and curriculum[13]. The purpose of the CUPM was to develop a broad mathematical program to improve the undergraduate curriculum in the United States' colleges and universities. The committee recognized the need for many more mathematical experts in education, government, and industry which needed to be addressed immediately[14]. They developed recommendations for high ability mathematics students to prepare them for research in mathematics at the university level[15]. This committee focused on three main groups of students: students who would continue with mathematics graduate work, students who took mathematics with the goal of applying it to science or engineering, and students who were planning to become elementary or secondary school teachers.

The CUPM felt that all teachers must have a solid understanding in traditional mathematics with an appreciation of modern mathematics[16]. CUPM recommended one year of algebra and one year of geometry in high school as well as at least twelve hours of college mathematics for future elementary teachers. The CUPM also stressed that at least 20% of elementary teachers needed to have stronger mathematics than this minimum. The CUPM recommended 28 hours of college mathematics to include at least three courses in calculus, and one course each in geometry, abstract algebra, and probability for teachers in middle school and high school who may teach mathematics. For high school teachers who taught mainly mathematics, the CUPM stated they should receive a bachelor's degree in mathematics consisting of at least 33 hours of college mathematics classes[17]. This would consist of higher level courses such as abstract algebra, analysis, linear algebra, probability and statistics, and advanced geometry. The CUPM also recommended that high school teachers who taught linear algebra, probability, or calculus should have a master's degree with at least two-thirds of their coursework in graduate level mathematics. The CUPM did mention that stronger content was needed, but pedagogy needed to be addressed as well. The teacher's understanding of mathematics affects his or her attitude, and the CUPM recognized the need for motivation and positive attitudes in mathematics for teachers as well as students. All of the above recommendations were made by mathematicians; many mathematics educators, however, did not feel that such a high level of mathematics was necessary. Many of the new college course textbooks that were published during the 1960s related to the mathematical content as established by CUPM. Nevertheless, the movement to increase the standard of teachers in mathematics was underway and continues today.

Pedagogy

     Teachers had to not only learn new mathematics content, but many had to be educated in new teaching techniques such as guided discovery learning. Discovery learning often means very different things to different people[18]. Kersh states that when students learn by discovery, they develop an interest in the task, they understand the material better and thus are better able to remember and transfer the learning, and they learn a strategy for discovering new things. However, the role of the teacher in "new math" programs was one of guided discovery—a process in which students discover principles with considerable help and guidance from the teacher. Teachers and the textbook guide the student to solving problems by asking pertinent questions at appropriate times[19]. Many educators and psychologists believe that guided discovery works well in teaching because when students are encouraged to think for themselves, they not only understand more, but it acts as a motivational force to generate enthusiasm and interest.

     Students learn during guided discovery by the practice that they do to solve a situation. Furthermore, drill can accompany guided discovery by offering practice to students once they understand the process[20]. Guided discovery is a nice balance between direct instruction and independent discovery learning because the student can benefit from both types of instruction at the most appropriate time. Guided discovery became the most popular teaching method during the "new math" era affecting all teachers teaching this new curriculum.

     However, the lack of pedagogy explanation and research in materials that reference the "new math" movement shows that pedagogy was seen as something to "take a look at" rather than an integral part of changing the curriculum. While many of the projects stated that they promoted discovery learning, very little was explained or promoted in the actual materials. Therefore, teachers knew they should use it, but they were not sure how to use it in their classrooms.

Aftermath

As the "new mathematics" movement continued, publishers began to label all new textbooks "modern" or "new" despite whether or not the book followed similar principles as the materials published by the many projects. By the mid-sixties, most new curricular materials were being commercially produced. The original philosophies of many of the projects writers were not present in some of these textbooks. The validity of the projects was questioned as the popularity was considered to come from the Hawthorne effect—an improvement occurs in a study because of the enthusiasm or increased effort of the participants. Changes in the "new math" projects were readily accepted at first; however, as time passed and results were not as positive as expected, critics stated that mathematics was in a state of massive confusion. One author described the "new math" era as "the over-all academic mayhem that is going on in the name of reform"[21, p9].

An explanation of the difference between the secondary movement and the elementary movement is necessary. Many of the "new math" materials were being used in the elementary grades without further education of the teachers. Elementary teachers (or their administrators) "jumped on the bandwagon" of the "new math" movement, which at the beginning of the reform, had concentrated much more on the secondary level. Furthermore, many elementary teachers who were forced to teach the "new math" in their schools did so with an incredibly large void of necessary mathematical knowledge. Much of their own training in mathematics quite possibly was minimal and very traditional. Also, many elementary school teachers do not favor mathematics, and introducing more abstract concepts only caused confusion for both the teacher and his or her students[22].

During the 1970s a strong attack on the "new math" reform movement was launched blaming the "new math" as a curriculum that produced children who could not add[23]. Parents were, in general, dissatisfied with their child's ability to "do mathematics." Many were also unhappy with the lack of accountability of the projects. Many programs were not tested but abandoned quickly during this backlash.

In fact, the pioneers of the "new math" movement were not as influential in the end because the "new math programs" that reached schools lacked the pedagogical ideas expressed by the pioneers.

Many different groups were involved in the "new math" movements, and their interests began to change as time evolved. As focus changed, much of the funding from private and federal sources began to cease. Conflicts arose among the original pioneers of mathematics curriculum reform as the "new math" movement developed on a nation-wide scale. Both Max Beberman of UICSM and Robert Davis of the Madison Project became disillusioned with the effects of the "new math" movement when their idealism was being challenged by schools and programs introducing "new math" ineffectively. Dr. Beberman warned that hasty promotion of the "new math" would cause problems, particularly in the absence of adequate teacher training. Beberman felt that the general public was being misguided concerning some new topics or content. For example, he felt that the idea of sets, an idea meant to be simple and helpful, was depicted as very confusing and domineering within the curriculum[24]. Beberman realized that we must learn from the projects and adjust them accordingly. By 1962, he stressed that more computational skills must be taught within the curriculum in elementary grades to have an effective high school experience in mathematics; also, he felt that the total content of the UICSM course may be too much for four years of study. Beberman was so disillusioned with the problems that developed, he went to England in 1971 to study mathematics reform there and died shortly after arriving[25].

Another member of UICSM cautioned the public and schools to slow down and examine various projects before implementing a program into a school. People had misunderstood the "new math" movement by assuming that no computational skills were encouraged—clarifying that the project writers were trying to achieve a balance between drill and reason. Fehr, a past president of NCTM, also began to speak against the "new mathematics" programs—specifically the elementary ones in which he felt not enough concrete mathematics was being taught[26].

The "new mathematics" movement began to die when those most supporting it began to reduce their efforts. Many of the mathematicians were concerned with "new mathematics" content as the sole purpose of implementing "modern mathematics" while others hoped that "modern mathematics" would improve teaching strategies and pedagogy. By 1965, major project writers had completed their curriculum materials, but the problem became teacher training and classroom implementation. While many educators, in retrospect, believe that the content, basic philosophy, and ideas were great, the implementation was terrible. Many criticisms formed as a result of the "new mathematics."

In 1962, an article including a letter criticizing the "new math" signed by sixty-five mathematicians and its rebuttal by Ed Begle (director of SMSG) appeared in the *Mathematics Teacher*. Some of their criticisms were that the project authors were expecting similar mathematics abilities in classrooms, some material was being introduced too early, and students were being exposed to new terms and concepts without a concrete preparation. Obviously, the tides were changing concerning general support for the "new math." However, Begle attempted to show that most of the challenging ideas expressed in the original letter were actually personified in the SMSG program. In other words, the "better" ideas expressed in the letter by many mathematicians were shown to actually be part of the SMSG project—making the mathematicians appear to be unclear about the actual philosophy and content of SMSG and other "new math" projects.

One major criticism was that no group seemed to answer the question of why the "new mathematics" was better than the traditional mathematics. Furthermore, many project authors did not analyze which areas of the traditional curriculum were ineffective before developing a new one[27]. Did "new mathematics" solve the problem of America's poor performance in mathematics? Some would say "no" as the United States was reported as having done the worst in the first International Study in Mathematics Education published in 1964[28]. Many project authors developed a new curriculum that pushed down some college mathematics into the secondary level, but most did this without making sure that the college material was necessary or appropriate. Some complaints were heard by science teachers annoyed when students in their own classes could not perform elementary mathematics functions in order to solve a science problem because they had not yet learned the topic because of the changed curriculum. Other educators were concerned that some of the projects leaders, particularly SMSG writers, were trying to develop a national curriculum—taking away the responsibility from the states. Therefore, there existed no overall coordination of the "new math" movement or the direction it should head to combat the uneasiness of those who wanted to know what the benefits of the "new math" were supposed to be.

Another major criticism of "new mathematics" was that the teachers were expected to understand and teach the new curriculum with ease, but in reality the teachers were not able to do either[29]. Some felt that a mathematics specialist should have been made available for all teachers in schools (and many would agree this would be an ideal situation today). Many elementary teachers who already had a weak basis of mathematics could not become experts in "new mathematics" in one summer institute. "Teacher training in the use of the curriculum was too little and too late"[30,p.193]. And, by the mid-sixties, many of the NSF institutes were no longer being offered because of funding problems. Teachers also were expected to teach the new curriculum without a choice (the decision being made elsewhere). Often the administrators of a building or district would make the decision about which "new math" program, if any, would be used in the school; leaving the teachers out of decisions affecting themselves directly. Without ownership in the decision, many teachers did not put forth the effort required for a successful program. "Most often, the teachers' philosophy did not mirror the spirit of the work of the movement"[31,p.181]. Many of the teachers who were involved in writing the "new math" materials were often teachers with very high abilities themselves who often taught high ability children. Therefore, the average teachers, particularly elementary teachers, had problems with these new materials. Many of the "new math" projects realized that teacher training would be an issue, but most had no idea of how important an issue it would become.

In general, better teaching was needed at all levels in order to improve the mathematics curriculum. The heart of the problem of the "new math" movement and the desire to improve curriculum was teacher training—both in-service and pre-service. The issue of teacher improvement did not receive the attention necessary to improve mathematics.

Another criticism was that some of the projects were developed mainly by mathematicians with little input from teachers. Many felt that parents and children, as well as teachers, were left out of the process of creating new materials[32]. Furthermore, the textbooks were sometimes written by mathematicians in an abstract manner with a lack of relationship to the real world, and without thorough knowledge of how younger students learn. In fact, many would say that much of the material produced, particularly by SMSG, did not have the same underlying philosophy or ideas of how children learned mathematics. Along with the criticism of abstraction, some felt that the stress on precise language created problems unnecessarily for students. However, mathematicians still advocate the use of precise language in a field that is naturally abstract in order to help make the topics clearer for students overall. Another complaint was that some of the projects produced materials too quickly without much revision. For example, the SMSG group wrote seven years of curriculum in one year; it was tried and tested during the next year; and then the materials were made available for nationwide use[33].

A major criticism of the "new mathematics" era was that many project writers focused mainly on students with high abilities in mathematics, and the reform needed to analyze teaching methodology and content for all academic levels of students[34]. Allendoerfer agreed that we were neglecting the non college-bound students while increasing all our effort to the college-bound. "We have taken a structure, which was well built to start with, have shifted its partitions, redecorated its rooms, put in new and faster elevators, dressed up the lobby, and cleaned the grime off the exterior. Certainly it is more livable that it used to be, but it is still intended for the use of the upper classes"[35,p.692].

All students were supposed to be able to answer "why" to mathematics questions rather than just being able to do mathematics. In reality, logic took precedence over learning the basics, and the majority of students in mathematics, many having average ability, were found to be lacking in basic procedures such as arithmetic. Critics such as Kline, stated that too much emphasis was being placed on set theory and precise language—so much so that students were learning words that had no meaning to them[36]. Kline further stated that teachers needed to relate mathematics to students' lives in order for it to make sense. In other words, logic in understanding would mean nothing if it was not related to the children.

While some of the project leaders developed materials aimed at low ability and average ability students, most did not until at least the mid 1960s. In 1964, NCTM proposed that a council should be established to investigate the problems of low ability students and underachieving students[37]. The problem that arose for low ability students was that they had difficulty when taught with materials developed for students with high abilities. The negative influence on these students would not allow the later materials produced specifically for lower ability students to be considered. One letter signed by a group of parents in 1958 to Max Beberman concerning the UICSM project being taught at the University High School showed that students who were struggling began to dislike mathematics and develop poor attitudes—an indication that a "backlash" to the "new math" movement was possible if steps were not taken to improve the negative attitudes some students began to develop[38].

It was not until later projects that the reform shifted from mainly the secondary school curriculum to the elementary school curriculum. While many thought it logical that new programs start with the elementary level, many reasons were given for the focus placed on secondary level instead. For example, one of the strongest forces behind the movement was improvement of mathematical ability of college-capable students as quickly as possible—thus starting in elementary school would not provide the quick response to the nation's desire to improve its mathematics and science ability. Many project leaders also felt that it would be easier to train secondary teachers to implement these new programs because there were fewer of them than elementary teachers, and most secondary mathematics teachers had more training in mathematics than elementary teachers. The project leaders felt that utilizing the secondary mathematics teachers would be an effective way to successfully incorporate these new programs. However, many educators would argue that it was finally realized that the work at the elementary school level needed to be addressed first as a basis for effective work at the secondary level.

"New mathematics" was encouraged for all students despite much support through research[39]. Those criticizing the assessment of the "new math" projects often condemn the lack of research of the project materials as well as the lack of appropriate assessment for students. First, the lack of research to support project materials was criticized. Bosse stated that the "new math" movement was based upon theory only and not empirical evidence[40]. Many educators called for more critical research pertaining to the content of the new curriculum as well as effects on students. Furthermore, too often the materials schools used were not adequately tested by the projects. For example, the SMSG materials were only piloted one year and then widely circulated. SMSG may have benefited more by testing periodically while increasing their circulation. Therefore, more research of the actual project materials was necessary, in both quantitative and qualitative forms, to provide rationale to the doubting members of the mathematical community that "new math" materials were valuable.

Furthermore, many of the projects were involved in their own formative evaluations—evaluations performed during trials and testing of materials. While this was important for the development of materials, summative evaluations would have been helpful and necessary to truly evaluate the projects of the "new math" movement.

It was apparent that the authors of the "new math" projects knew that experimental research was necessary; but it seems that they were not sure how to assess the validity of their projects[41]. In order to do so, assessments of students and their abilities to understand mathematics were needed. The leaders of the projects knew that research was needed to assess not only the computational skills, but also the level of abstraction students achieved. Many educators, both supporters and critics, believed that the "new math" movement was never accurately tested and that part of its perceived failure should be placed on the assessments given rather than on the movement itself.

While some projects researched the use of their own materials pertaining to student knowledge of mathematics, most did so in terms of standardized testing. The majority of assessment tools such as the SAT (Scholastic Aptitude Test) did not necessarily follow the goals of the "new math" movement[42]. The "new math" movement taught understanding of mathematics and taught children how to reason and explain their thought processes. Tests needed to be created to assess this reasoning, and the traditional standardized tests did not do so. This created the problem of testing the materials because without proper tests and specific goals, it would be difficult to see if students developed their abstract thinking and problem solving skills. Therefore, the student mathematics assessments used were never in line with the goals and objectives of the projects[43].

However, comparing different curricula through testing was seen as impossible. Because each curriculum had different goals, a test aimed at one set of goals would surely favor that specific curriculum being tested over another. Therefore, the "new math" could not be "worse" or "better" than the traditional curriculum because there was no single way to compare the two movements. The differences could be described, but testing the effects on mathematics learning would be much more difficult. Therefore, evaluation of the programs would be difficult because of different objectives causing different outcomes (both mathematical and pedagogical). While this is not impossible, the leaders of the "new math" movements were unsure of how to accurately test their materials against the traditional ones. Since standardized testing and solving problems were traditionally the way students were assessed, the leaders of the "new math" projects possibly did not understand how to implement other means of assessment that could show if students truly understood and explained their mathematical reasoning.

Robert B. Davis stated that many people did not know what "new mathematics" really was—they only knew that it was being adopted throughout schools in the US. But was it? It has been suggested that in most classrooms the "new math" was never really tried[44]. Davis explained that a very large number of schools did not alter their curriculum at all during the "new math" movement. Another large segment of schools thought they were adopting "new math" ideas because their new textbooks used words like discovery and mathematics structure. However, the only section of the population who truly understood what comprised the "new mathematics" was the third population: a significant minority of schools who genuinely taught a "new mathematics" program[45]. Davis described a classroom which understood the "new math" program as one where definitions were clearer, student participation was greater, and teacher comprehension was adequate in both mathematics and pedagogy. In such a classroom, the mathematics curriculum would have new material, old material presented in a new way, or some deleted material no longer needed. Most of all, a creative atmosphere would exist in the mathematics classroom.

"New mathematics" promoted understanding rather than rote learning, but it did not solve the problem of excessive rote teaching or learning in our schools. While many schools continued with the traditional teaching of mathematics, those most affected by "new math" were small cities and suburban areas[46]. Many rural areas and small-town schools never adopted the "new math," and those larger schools who did adopt it often did so only for the college-bound students while still offering the traditional curriculum for all other students[47].

While mathematics educators attempted to change the mathematics curriculum on a national scale, only those in the educational community assisted in the extension of this reform movement. One problem in this is that often parents and administrators were not properly informed of the purpose or the benefits of the program. And, while the mathematics educators were doing a fine job with the curriculum, they were not examining social pressures, effects on the public's attitude about the Vietnam War, the peace movement or civil rights.   Many of the general public felt that the goal of mathematics reform was completed by the mid-sixties, and their focus was changing from mathematics and science education to social education. Therefore, the projects did not have the public's full support in implementing a program—support that would have been necessary to accomplish the goal of establishing an improved "new math" curriculum in the country's schools. While the "new math" curriculum had merit, the negative effects of implementing the "new math" program in some of the country's schools caused a backlash to the movement. Focusing on teaching and improving training rather than content could possibly have been a solution to improving the mathematics standards of our students.

Reform From the Mid 1960s to the Present
        The influence of "new mathematics" is present in the current educational requirements and curricula, but before discussing the current mathematics reform, a brief analysis of mathematics reform movements between the end of the "new math" and present day is necessary.
        "The 'new math' did not survive the 1960s"[48,p.240]. By the early 1970s, Max Beberman, the influential UICSM leader, had died, and SMSG was losing momentum. The late 1960s and early 1970s saw the introduction of other mathematics projects different enough that they were not considered part of the "new mathematics" movement. The early 1970s brought a backlash to the "new mathematics" and a renewed emphasis on paper and pencil skills.   President Nixon published a document in March of 1970 titled *Education for the 1970s: Renewal and Reform*.   In this document, the President called for a national reform in education particularly concerning education for the poor; financial planning of schools; and enhancement of education that takes place outside of school[49]. While the disadvantaged child was still not receiving an adequate education, Nixon stated that we know now that the socio-economic environment of a child affects the child more than the quality of the school he or she attends.   Nixon, therefore, called for a change in focus from mathematics and science to early childhood education, literacy, and equal education throughout the United States. Nixon further stated that the United States needed a focus in education based on experimentation and research.

Nixon also focused on the improvement of higher education, and this affected K-12 schools because it supposed that more students would seek higher education requiring a more advanced and challenging school experience. Nixon proposed to revamp student aid allowing all students who wanted to attend college the opportunity to do so. He also encouraged students to attend community colleges. Despite which type of further education a student chose, Nixon's proposal was to make that dream a reality with support to both students and institutions. Therefore, Nixon hoped to achieve equal education opportunity for all in higher education as well. As can be seen through Nixon's proposals, emphasis was changing from a scientific one to a social one throughout the entire K-16 school system. Furthermore, Nixon called for accountability within the school systems. By expecting accountability, the nation called on schools to show their students' progress. Traditional standardized testing was the easiest way for schools to show the content their students had learned in short periods of time. Reverting back to traditional mathematics allowed for traditional testing, and this offered a way for the public to see the school's accountability (measurability) [50].

As the mathematics curriculum focused on a back to basics reform in response to the overreaction to and disfavor of the "new math" movement, the National Council of Teachers of Mathematics, saw the direction for mathematics curriculum across the nation disintegrate during the 1970s. Therefore, the Conference Board of the Mathematical Sciences National Advisory Committee on Mathematical Education (NACOME) of NCTM published the *Overview and Analysis of School Mathematics Grades K-12* in 1975 in which they began to discuss where the future of mathematics should be. Mathematics seemed to be in a chaotic state during the 1970s as teachers, parents, and administrators who were struggling with the "new math" quickly reverted back to the traditional, familiar style of teaching based on skill, drill, and practice. Because the United States lacked a national curriculum, modern curriculum reform was confusing to many educators.

As previously stated, the 1970s became known as a "Back to the Basics" movement. Because of declining test scores, teachers began focusing more on computational skill in algebra and arithmetic. However, NACOME declared that there were major problems in current practices of evaluating mathematics education." Many of these tests were not sensitive to the specific objectives of programs, and many were declared culturally biased—the tests were written and designed with the dominant culture in mind without regard to language or cultural differences of many minority students.

By the mid 1970s, NCTM could see that rather than learning from the "new math" movement and altering goals to proceed from this point, many educators had retreated to traditional drill and practice methods of teaching mathematics. A general view was that educators had to be either "for" or "against" the "new math". Thus they had to like the "new" as opposed to the "old" or emphasize "concepts" as opposed to "skills." NCTM saw this as unnecessary battling because educators should be learning from the past and applying the benefits of what they have learned rather than being involved in a war of "for or against." They warned of over-emphasis on basic skills, and stated that basic skills could be applied easier as many calculators came into the classroom. They also saw the emergence of many different issues that would become apparent in the 1980s such as mathematics for students with limited mathematical ability, the changing influence of computers and calculators, and the increase in statistics and probability topics into the curriculum[51].

Mathematics curriculum also began to be dominated by state departments of education in the 1970s and 1980s. Therefore, rather than allow reform to happen and acknowledge once it began, the National Council of Teachers of Mathematics decided to lead mathematical reform in the United States. The first time NCTM took control of mathematics reform was with the publication of a document entitled *Agenda for Action* in which NCTM stressed the need for problem solving during the 1980s. Thus, the "Back to Basics" movement in mathematics switched again to more "critical thinking."

The following recommendation were made in the *Agenda for Action*[52]:

1. Problem solving should be the focus of school mathematics.
2. Basic skills must be taught more than computational ability.
3. Calculators and computers should be used in conjunction with learning mathematics.
4. The teaching of mathematics must be efficient and effective.

5. Assessment of understanding should extend beyond traditional testing.
6. Mathematics should be encouraged for all with options for differing abilities and needs.
7. Mathematics teachers should have a high degree of professionalism.
8. The public should understand and support the need for mathematics.

This agenda is considered the forerunner of the *Standards* and *Principles and Standards for School Mathematics*, to be discussed in the next chapter. Many of the ideas expressed in this agenda were encouraged in the 1980s, but had a larger influence in the *Standards* which affected mathematics curriculum in the 1990s. Other issues promoted in the agenda were: teacher motivation of students; logical reasoning and precise language; guided discovery and group work techniques; increased years of required mathematics to graduate; and high standards of teacher training.

The PRISM report (Priorities in School Mathematics) was published by NCTM in 1981, and this report discussed findings from a national survey completed by members in education as well as the general public. The authors of the report discussed teacher input and thoughts concerning the eight principles from the *Agenda for Action* that was published a year earlier. In summary, members of society as well as educators were surveyed and stated the following: clearly there was a need for more problem solving in school mathematics; what encompasses basic skills is unclear to many educators and thus, teachers do not know which areas to emphasize; teachers were aware that computers and calculators needed increased emphasis; teachers wanted and would use more variety in teaching if supplied with resources and instructional strategies; teachers strongly felt that evaluation and assessment needed addressing; if new courses were added to the curriculum, support would have to be developed; and improved pre-service and in-service education of teachers was necessary[53].

Another major report that increased the public's interest in education was *A Nation at Risk,* published in 1983 by the National Commission on Excellence in Education[54]. This report declared the United States as a nation at risk because of our lack in strong education. It "awoke a sleeping nation to alarming problems in our educational system"[55,p.2]. This report helped to establish a climate that would support educational change.

In 1989, another mathematics education book was published shortly before the *Standards*. *Everybody Counts* was published by the Mathematical Science Education Board of the National Research Council. *Everybody Counts* was a book written for the general public that increased knowledge of the importance of mathematics. The authors of this book also spoke of the need for effective teaching and professionalism among teachers. In *Everybody Counts*, the need for teachers to lecture less and adopt a more student-centered classroom where problem solving is emphasized was explained. Understanding of mathematics was expected to be the most important concept. The authors also described various issues about mathematics such as the importance of technology, research, and demographics. This publication stressed that mathematics provided knowledge that would allow citizens and the United States as a nation to compete in a technological economy. However, the most important issue promoted was change in education to improve the way students learned mathematics. Many believe that this publication added to the success of the *Standards*[56].

The *Standards* was launched by NCTM in 1989, and this publication dominated the mathematics education community as it began to lead mathematics reform in the United States. Therefore, the largest driving force for mathematics reform in the 1990s was the National Council of Teachers of Mathematics and its publication of *Curriculum and Evaluation Standards for School Mathematics*. This book was followed by a second edition called *Principles and Standards for School Mathematics* in April 2000. Both books have continued to lead a national mathematics community with guidance that constitutes current reform. The next chapter deals with this current mathematical reform and the similarities between it and the "new math" movement.

Connections with the *Standards*

The similarities between the "new mathematics" reform movement and the present mathematical reform movement will be addressed in this chapter. The connection between these two movements is based on the 1989 and 2000 publications of the National Council of Teachers of Mathematics *Standards*—generally agreed to be the basis for the current reform movement. Therefore, the goals of the *Standards* as well as the content of the *Standards* will be explained. Further, the current mathematical curriculum and the "new math" movement will be compared concerning philosophy, teaching, content, pedagogy, and assessment. Many of these characteristics will be similar to characteristics of the "new mathematics" movement.

The Standards

"The broad impact of the NCTM *Standards*, especially the *Curriculum and Evaluation* Standards, has been widely acknowledged within mathematics education, and by authors who come from outside of mathematics education as well"[1,p.81].

On March 21, 1989, the National Council of Teachers of Mathematics released its *Curriculum and Evaluation Standards for School Mathematics*, after five years of development. The While NCTM had offered recommendations in the past, the launch of this book was the beginning of a large, national initiative to influence education. In April 2000, NCTM launched the second edition of the *Standards* called *Principles and Standards for School Mathematics*. The content and theory is consistent in both editions; the major difference in the second is the format and layout of the information designed to be more concise and easier to understand and apply.

Furthermore, with the advance in technology, the new *Standards* utilizes resources via the internet and a CD.

NCTM had a tradition of taking a secondary role in curriculum revision. NCTM finally saw its role as a leader if the members of the organization united to speak out to influence and improve educational policy for mathematics teaching and learning. NCTM launched the *Standards* with the help of a public relations firm. NCTM promoted this curriculum philosophy by sending information to most school districts or administrators throughout the country—much like the CEEB did during the "new math" era to promote its curriculum philosophy.

The *Standards* defines an educational philosophy more than dictates a curriculum. The *Standards* begins with a discussion of societal and student goals as "Principles for School Mathematics" and then outlines the content that should dominate mathematics curriculum for grades Prekindergarten–12[2]. It also addresses new topics such as technology and mathematical modeling which represent problem solving skills in "real life" situations—often through the use of discovery learning.

The sections of the current *Standards* are split into four groups: Grades Pre-K–2, Grades 3–5, Grades 6-8 and Grades 9–12. In each of these groups, the first five standards are content standards with the following topics: Number and Operations, Algebra, Geometry, Measurement, and Data Analysis and Probability. Within each grade level group, content standards are discussed with respect to specific mathematics topics that should be taught at that level. These include geometry, whole number computation and operations, number sense, estimation, measurement, fractions and decimals, patterns and relationships, and statistics and probability in grades Pre-K-2 and grades 3-5. Topics for grades 6–8 include algebra, patterns and functions, probability and statistics, geometry, computation and estimation, measurement, mathematics structure, and number. Topics for high school include algebra, functions, geometry, trigonometry, statistics and probability, calculus, and mathematics structure. While many believe that the abstraction of concepts characteristic of the "new mathematics" era is absent from the present curriculum, some of these concepts are included throughout the curriculum, but not as abstractly as before. The inclusion of a variety of topics such as discrete mathematics, patterns and functions, estimation, probability and statistics, and earlier introduction of topics such as algebra and geometry shows the influence of the "new mathematics" program in deviating from the traditional curriculum and including new mathematics as well as "old" mathematics with a more applicable and understandable meaning.

The next five standards in each group are called process standards, and deal with Problem Solving, Reasoning and Proof, Communication, Connections, and Representation. Communication deals with the students being able to communicate effectively about mathematics and while using mathematics. Reasoning and proof and problem solving require the student to be able to understand a problem, reason through the problem using critical thinking skills, and arrive at a solution that the student can assess. These characteristics should be applied in every situation. Connections deals with a student being able to make and see connections within mathematics among topics such as geometry, arithmetic, etc. as well as between mathematics and other subjects in school and issues outside of school. Representation asks that student be able to represent and solve problems in a variety of ways.

Besides these specific mathematics topics described above, *all* students are expected to achieve the six following goals: 1) learn the value of mathematics (similar to application to every day life); 2) become confident in one's own ability (mathematics for all); 3) become a problem solver (critical thinking); 4) learn to communicate mathematically; 5) learn to reason mathematically (understand the concept); and 6) make mathematical connections (applications to the real world) . These goals directly relate to the ideas behind the "new mathematics" movement of teaching children to understand mathematics by increasing critical thinking and problem solving skills.

NCTM continued its work in mathematics reform by publishing the *Professional Standards for Teaching Mathematics* in 1991 and publishing the *Assessment Standards for School Mathematics* in 1995 before the 2000 publication of *Principles and Standards for School Mathematics.*

The *Standards* had a national influence because of its influence on other professional organizations and states. The *Standards* influenced other professional organizations because it served as a model of how to develop national standards without government influence[3]. In fact, many other organizations' standards have similar format and philosophy as set forth by NCTM. Standards now exist for science, history, English, physical education, art, economics and geography education among others.

The *Standards* has also served as a model for many states who have and are developing their own state curriculum standards. Some states were developing their own curricula at the time the *Standards* was published, and many felt the publication of the *Standards* supported the state since the goals of each were similar. In fact, a total of forty-nine states now have some mathematics standards curriculum established—many of them influenced by the *Standards*[4]. While it is not known exactly how many teachers or schools are following the ideas as established in the *Standards*, NCTM realizes that change occurs slowly, and they must provide enough time to see that change.

As mentioned above, the National Council of Teachers of Mathematics has released an updated version of the *Standards* entitled *Principles and Standards for School Mathematics*. This new edition was launched at the NCTM annual conference in April 2000[5]. This edition is based upon the foundation set in the original *Standards;* it integrates curriculum and evaluation standards, professional standards, and assessment standards into one document. Based on the previous publication, the new *Standards* serves as a guide for mathematics teachers incorporating more technology while maintaining the high standard of teaching and mathematics curriculum guidelines. NCTM felt the need to publish an updated version of the Standards because they felt that mathematics reform should constantly be improved based on teacher feedback and the changing world[6].

Clemens correlates the 2000 *Standards* with the "new math" by explaining, "the last time the bar was raised too high, the result was the so-called 'new math', which cannot be judged a success by any measure"[7, p.1074]. Therefore, while the *Standards* has been successful, educators must proceed with caution as they are implemented to a larger degree to achieve positive mathematical reform—to avoid the frustration and confusion the "new mathematics" movement brought for most. Furthermore, some educators hope that the *Standards* will be truly implemented throughout the country, and that the public does not perceive them early on as a failed reform movement. In other words, the *Standards* must be given a fair chance for widespread implementation before the public decries their failure quickly as happened in the "new math" movement.

The comparison the "new math" movement with the current movement is based on both publications of the *Standards*. Because the second edition has only minor changes to the first, the educational reform that has taken place in schools in the past ten years and is taking place today is a result of the both the original and updated *Standards*.

Historical context

Many saw the "new math" movement as a reaction to the launch of *Sputnik* as well as the CEEB's report. In the same way, some may say that the *Standards* was a reaction to the report, *A Nation at Risk*. However, the *Standards* was not formed as a reaction to this report, but instead was a gradual evolvement of the NACOME report and an *Agenda for Action*[8]. The public's interest in it most likely came from the increased fear of America's students once again doing poorly compared to foreign students[9]. This continued with the increase in the number of international comparisons constantly showing American children low in mathematics ability[10]. Furthermore, with the conflict surrounding the Gulf War, America was forced to look at its own technology through mathematics and science education. The nation felt threatened again, and public interest helped support and promote the *Standards*.

In order to compare the "new mathematics" movement with the *Standards* movement, conclusions of the "new math" era overall will be contrasted with the similarities and differences of the *Standards*. To provide this information in an easier context, the similarities and differences will be based on the following categories: philosophy, pedagogy, teachers, content of the mathematics curriculum, and assessment.

Philosophical Similarities and Differences

First, the philosophies of the "new math" movement and the *Standards* movement will be discussed (see Table 1). Probably the most important underlying philosophy of the *Standards* that corresponds directly with the "new mathematics" movement is the overall goal that students should *understand mathematics*. "The experience of problem solving is the means to that end. That experience should be fixed in student recollection as definition of what mathematics learning is"[11,p.91]. Besides problem solving, emphasis on all three process strands of connections, reasoning, and communication in mathematics help to achieve the goal of understanding mathematics.

Table 1.

The Philosophies of the "New Math" Movement Compared to the Philosophies of the *Standards* Movement

"New Math"	*Standards*
The main philosophy of the "new math" movement was that children should understand mathematics by actively participating in their learning.	The major philosophy of the *Standards* is that mathematics must be studied by all children with a goal of understanding.
The many "new math" programs led a reform that was national in scope in order to improve mathematics standards and curriculum throughout the nation's schools.	The *Standards* is a movement that is national in scope and led by NCTM. It is supported by many states through their own mathematics standards.
All programs began a step to bring students "up to date" with a modern curriculum that would reflect the more technological and modern world in which they were living.	The *Standards* reflects that technology has changed the world and must be incorporated appropriately into the mathematics curriculum.
The "new math" movement saw the need for modernization of mathematics particularly for the advancing technology that was evident in World War II and the Cold War.	The current mathematical reform has been led by the National Council of Teachers of Mathematics and has been in response to a need to develop a more mathematically able student for a more technological society.
The "new math" movement sought to integrate the mathematics curriculum in order for students to make connections among mathematical topics and between subjects, particularly science and mathematics.	The *Standards* promotes integration of mathematics topics among all mathematics courses. In particular, the *Standards* supports connecting mathematics with all subject areas.
The "new math" projects began with an intention of developing materials for higher ability students, but then moved toward	The *Standards* began with the philosophy of "mathematics for all."

creating textbooks and materials that could serve all ability levels.	

The *Standards* recognized the need for all students to learn mathematics in order to prepare students to become productive citizens. In order to do this, the *Standards* tried to offer traditional mathematics content such as algebra, geometry, and trigonometry with newer topics such as probability and statistics (important in the "new math" era also) and discrete mathematics[12]. Memorizing isolated facts and performing manipulations has given way to conceptual understanding of mathematics. In order to achieve this, students are actively engaged in mathematics by exploring mathematics in cooperative groups and using techniques associated with discovery learning. The *Standards* has also stressed the importance of using technology when appropriate.

Probably the most important similarity between the "new math" movement and the *Standards* movement was the mutual desire to improve mathematics education in schools in an advancing technological society. One of the major influences in writing the *Standards* was the changing technology in the world and the desire to incorporate it into the mathematics curriculum. We see a demand for students to take more mathematics courses because of the importance of mathematics and science to our nation. The "new mathematics" movement stressed the importance of developing science and mathematics skills in students because of the new technology affecting the United States. As seen in the "new math" movement, one of the issues affecting the *Standards* was the need to update mathematical content in our mathematics curriculum.

Because society was becoming more technological in the 1980s as well as the 1960s, both the "new math" movement and the *Standards* movement saw the need to integrate the mathematics curriculum. By doing so, students should see the connections among various mathematics topics as well as the connections to other subject areas. The "new math" movement focused on connections within mathematics as well as the connections between mathematics and science. The *Standards* movement has focused more on connections within mathematics and connections between mathematics and all subject areas.

The "new math" movement was criticized for heavily focusing on the college-bound students while ignoring the average ability and low ability students. The *Standards* on the other hand, encourages "mathematics for all." The "new mathematics" project authors did state that mathematics for all students needed to be addressed, but they began by focusing on college-bound students, and many project writers never developed materials for other ability groups (many because of the sudden abandonment of this reform era). The idea of mathematics for all was evident in most projects, but was not implemented successfully. While the *Standards* is seen as promoting mathematics for all, it also discusses the need to challenge high ability mathematics students. Therefore, while the initial focus of both reform movements was different, both reform movements have also encompassed mathematics education for all students despite mathematical ability.

Pedagogical Similarities and Difference
    Second, the pedagogical techniques of the "new math" movement and the *Standards* movement will be discussed (see Table 2). The *Standards* also promotes the use of guided discovery as a teaching method—relating directly to the "new mathematics" movement that also favored discovery learning. Cooperative learning is emphasized in both the *Standards* and the "new math" era. While it may have not been labeled that way, the idea of children working together rather than always in isolation was encouraged. Even though students are encouraged to work together, competition between schools is prevalent. Instead of competition among students in the classroom, competition has moved into the formation of school "math teams" that participate in competitions such as the American Mathematics Competitions sponsored by the Mathematical Association of America or similar competitions sponsored by the National Council of Teachers of Mathematics.

    The *Standards* promotes students "doing" rather than "knowing;" ideas should originate with the students rather than be dominated by the teacher; the teacher should stop lecturing and begin facilitating student discovery; and students should move away from routine drill and practice and learn more problem solving skills. These are all statements that are similar to ideas expressed in the "new math" movement. Students should understand mathematics through discovery learning with less drill and practice and more problem solving.

    The *Standards* is often associated with a constructivist approach for teaching[13]. The roots of constructivism are normally considered to lie with Jean Piaget in his beliefs that a student incorporates new experience into an already existing structure and then

Table 2.
The Pedagogical Techniques of the "New Math" Movement Compared to the Pedagogical Techniques of the *Standards* Movement

"New Math"	*Standards*
The most popular form of pedagogy discussed throughout the "new math" projects was the method of discovery learning.	The major pedagogical philosophy of the *Standards* is constructivism—a philosophy of teaching where techniques such as the discovery method, guided practice, and cooperative learning are often used.
The many "new math" programs promoted active learning by the students as they constructed meaning for themselves—often working in groups rather than individually.	The *Standards* promotes active learning through its constructivist philosophy where students are able to construct meaning for themselves in a student-centered classroom.
Most programs promoted the idea of the teacher as a facilitator rather than the authority while teaching.	The *Standards* requires the teacher to guide students in their learning rather than to rely only on direct instruction.
The "new math" projects often produced materials that they thought could be "teacher proof" in terms of providing resources that should easily incorporate new pedagogy; however, this did not work well and actual teacher training was needed.	The *Standards* and NCTM realize and support teacher training in how to use different pedagogical techniques. NCTM recognizes that the success of the *Standards* lies with the ability of the teachers to use the materials successfully.
The "new math" movement promoted active student involvement—something that had not been seen in the traditional curriculum.	The *Standards* promotes a student-centered classroom where students are experiencing mathematics with "hands-on" activities and the use of multiple manipulatives.

reorganizes those structures to understand new experiences. Many educators today would consider constructivism a teaching philosophy that normally encourages students to construct meaning through their own actions and learning. While it is considered a philosophy rather than a teaching theory, it often encourages techniques such as discovery learning and cooperative learning in child-centered classrooms. Overall, it encourages active learning by the students. This is the same underling philosophy prevalent in the "new math" curriculum materials as evidenced in the UICSM (1951) and Madison Project (1957) materials.

Concerning pedagogy, the Addenda Series and Navigation series published by NCTM to supplement the *Standards,* asks teachers to guide students in discovery and abandon the traditional teacher authority role[14]. Instead, the teacher must become a facilitator offering questions and guidance when necessary. While manipulatives can be used in the discovery method, teachers are encouraged to offer guidance with them when necessary. Teachers must also support mathematics by creating a positive learning environment for students, particularly at the elementary level so as not to discourage further study of mathematics[15]. Some of the books in the Addenda Series and Navigation series explain classroom situations, but many center ideas around activities offered, often with worksheets. In general, the authors have provided examples of problems based on the *Standards* detailing the content and pedagogy necessary.

NCTM, like the mathematicians leading the "new math" movement, started its focus on content rather than pedagogy because they felt people in mathematics education would find change in content easier to discuss than change in pedagogy. However, NCTM saw the importance of teaching techniques and pedagogy and published *Professional Standards for Teaching Mathematics* (often referred to as the *Teaching Standards*) in 1991[16]. This guide provided standards for teaching, for professional development, and for the evaluation of teaching.

The authors of the "new math" materials recognized the need for teacher training in the use of materials, some tried to develop "teacher proof" guides—ignoring the actual teacher training (both pre-service and in-service) that was inevitable for success. NCTM had learned from the "new math" movement that focusing mainly on curriculum change would not be sufficient. With another teacher shortage (as in the 1960s), NCTM realized that a strong guide for teaching standards and pedagogy would be necessary to improve mathematics. NCTM has been involved in teacher training by providing various supplements such as the Addenda and Navigation Series; local, regional, and annual conferences; and a web site that can be extremely useful for classroom teachers with little time or monetary resources. As with the writers of the "new math" projects, many saw the importance of not only changing content, but also of improving teacher in-service training for a more modern curriculum to be effective.

Teaching Standards Similarities and Differences
          Third, the teaching standards of the "new math" movement and the *Standards* movement will be discussed (see Table 3). There is a lack of mathematics specialists and mathematics teachers today as there was in the 1950s. During the 1960s, mathematics was considered a minimal subject for many high school students who took only one year of mathematics to graduate. Today, mathematics is encouraged for all; however, many states do not regulate the level of mathematics a student must achieve before graduation. They may require a specific number of years of mathematics a student must complete in order to graduate, but they do not stipulate what mathematics level students should reach.
          While teaching standards have improved, the lack of qualified mathematics teachers has caused some districts to hire teachers neither qualified, nor capable of teaching mathematics as was seen in the 1950s and 1960s as well. Problems surrounding this are the lack of emphasis on mathematics in the schools producing citizens with poor abilities in mathematics; the lack of qualified teachers; the problem of teachers leaving teaching very early in their career; and minimal high school graduation requirements.
          Furthermore, NCTM and various states realize the importance of pre-service education where prospective teachers need a solid background in mathematics and strong mathematics methods courses[17]. Requirements for certification and graduation from college have increased—particularly for high school mathematics teachers. For example, pre-service teachers who major in secondary mathematics education (in one Illinois college program) are required to have at least thirty-six hours of mathematics at university level, and approximately

Table 3.
Teaching Standards of the "New Math" Movement Compared to Teaching Standards of the *Standards* Movement

"New Math"	*Standards*
During the start of the "new math" era, there was a teacher shortage particularly in mathematics.	Today a teacher shortage in mathematics exists.
The many "new math" programs required practicing teachers to take courses to increase their mathematics content and required teacher education programs to increase standards for acceptance into the program and certification.	The *Standards* expects that certified teachers will have a solid background in mathematics. Furthermore, many of the increases in requirements for admission to college and teacher education programs as well as certification requirements are based on recommendations made in the "new math" era.
Many of the "new math" program materials were revised using teacher feedback from those teachers who were trained and who implemented particular project materials.	The *Standards* incorporated teachers in not only the writing of the *Standards*, but also in providing evaluations used to improve the supplementary materials and the 2000 *Standards*.
Most of the "new math" projects were dominated by mathematicians.	The *Standards* were written and led mainly by mathematics educators.

twenty-six of these hours are beyond calculus[18]. One state regulates that all secondary pre-service mathematics teachers complete thirty hours of mathematics with at least twenty-two beyond calculus[19]. Pre-service teachers obtaining a teacher education minor in mathematics are required to take at least fourteen hours of university level mathematics beyond calculus. Middle school pre-service teachers are required to complete a minimum of twenty-one hours of mathematics.

The National Council for the Accreditation of Teacher Education (NCATE) guidelines, based on NCTM and ACEI (The Association for Childhood Education International) guidelines for pre-service teachers, state that all prospective secondary mathematics teachers should have completed four years of high school mathematics including pre-calculus[20]. The NCATE and NCTM guidelines state that elementary and middle school teachers who specialize in mathematics should have a high school background including three years of mathematics that included two years of algebra and one year of geometry. Elementary teachers not specializing in mathematics should have a high school background of three years of mathematics consisting of an elective, algebra, and geometry. During college, elementary teachers must take a total of five hours (fifteen semester hours if they want to specialize in mathematics) in mathematics and mathematics education[21]. Middle school teachers must take twenty-four hours of mathematics and mathematics education courses. High school teachers must obtain the equivalent of a college major in mathematics for certification. Obviously, these requirements show an increase in the level and standard of mathematics required by teachers—a recommendation made in light of the "new mathematics" movement as organizations such as the Committee on the Undergraduate Program in Mathematics (CUPM) advocated for more hours for certification and graduation.

Teachers became involved in both movements because of the input of ideas and feedback into the materials published by each movement. The "new math" movement included programs that developed materials, tested them with teachers in schools, and then changed the materials based on feedback from the teachers. Likewise, NCTM widely distributed drafts of the *Standards* to its members and asked for feedback[22]. NCTM also did this with the development of the

2000 *Standards* and any other materials produced to complement the *Standards*.

Like the "new math" movement, the *Standards* was designed with collaboration among teachers, mathematicians, administrators, and mathematics educators. While the "new math" movement has been criticized as being dominated by mathematicians, the writers of the *Standards*, while emphasizing that much teacher input was used, were mainly mathematics educators at the university level[23]. Therefore, while both utilized teachers, the *Standards* have been led by mathematics educators rather than mathematicians.

Content Similarities and Differences of the Mathematics Curriculum

Fourth, the mathematics content of the "new math" movement and the *Standards* movement will be discussed (see Table 4). While many felt "new mathematics" faded away in the early 1970s due to its failure, no evidence supports this, and a closer look will show that many aspects of the "new mathematics" reform are present in today's mathematics curricula. While some topics may have decreased in emphasis such as set theory, many still exist in textbooks, and some, such as probability, statistics and problem solving appear more often[24]. Rasch stated that it was evident in her research that modern mathematics topics had not disappeared from the curriculum.

Requirements for graduation and entrance to college have increased for students. As can be seen in our current high school, college, and state requirements, the suggestions made in the 1960s forced change toward a higher standard in mathematics for all students. For example, many colleges currently require at least three years of high school mathematics, with some requiring four years for entrance into many college programs.

Besides raising the standard and level of mathematics, the "new mathematics" movement affected the structure of mathematics courses taught in schools. For example, the "new mathematics" movement called for a "pushed-down mathematics" curriculum where higher level mathematics was taught at an earlier level. The traditional curriculum before the "new mathematics" movement consisted of arithmetic taught in grades 1–8, algebra in grade 9, geometry in grade 10, and more algebra or geometry and trigonometry in grades 11 and 12. Since the "new mathematics" movement ended, the elementary mathematics curriculum has changed to include beginning aspects of

Table 4.

Content of the "New Math" Movement Compared to Content of the *Standards* Movement

"New Math"	*Standards*
The "new math" era called for a more modern curriculum. Much of the curriculum was redesigned with traditional topics "pushed down" into earlier grades than before.	Much of the content explained in the *Standards* is similar to that which was "pushed down" into the curriculum during the "new math" era.
The CEEB and other "new math" projects promoted the mathematical course outline for all high schools to include 1 year of algebra, followed by 1 year of geometry, followed by 1 year of advanced algebra and trigonometry, followed by 1 year of additional mathematics including topics such as calculus, probability and statistics, analytical geometry, or functional analysis.	This "pushed down" curriculum led to the creation of today's standard curriculum in high school: algebra, geometry, advanced algebra and trigonometry, and calculus or functional analysis.
Besides reorganization of the curriculum, some of the "new math" curriculum included new topics. While these topics existed, they had not been found in the school mathematics curriculum before. This includes probability and statistics, patterns, and logic/problem solving.	The *Standards* supports many of the new topics first expressed in the "new math" curriculum. For example, the *Standards* calls for the inclusion of probability and statistics, problem solving, mathematical reasoning, and measurement.
Many of the "new math" programs called for unified mathematics throughout the programs in order that students connect mathematics to other	The *Standards* promotes understanding of mathematics by requiring students to be able to connect mathematics within mathematics topics, across

subjects and other types of mathematics.	disciplines, and to everyday life.
Most of the "new math" programs developed curricular materials in the form of supplementary guides or textbooks in order that this new curriculum be implemented in the schools.	NCTM published an Addenda Series besides the *Standards* to help teachers understand how to incorporate this new curriculum into the classroom.
The "new math" movement focused on precision in language to help with understanding.	The *Standards* promotes understanding of mathematics by requiring students to be able to communicate mathematically.
The "new math" projects produced materials that caused teacher and students to move away from computation (drill and practice only) to more understanding of concepts.	The *Standards* promotes less focus on computational fluency and more on understanding than traditional curriculums. NCTM states that both are necessary for students to be successful, but traditional drill and practice along will not suffice.

algebra and geometry as early as first grade with the first formal course of algebra taught in 8th or 9th grade. Geometry is now evident in the elementary school curriculum. Also, the recently published *Standards* encourages *all* students to obtain a strong background in algebra by introducing algebra in Kindergarten, increasing in difficulty as students approach eighth grade; furthermore, students should be introduced to topics such as probability and statistics. This idea supports the "new math" curriculum of "pushing down" some traditional high school mathematics because students are capable of learning higher level mathematics at an earlier age.

However, many believe that the "pushed down" curriculum has not been "pushed down" enough. Davis of the Madison Project stated that we have an uneven pace of mathematics education. Grades 1–9 experience mathematics taught at a slow pace and slow advancement. However, grades 10–12 teach a majority of mathematics quickly for those college-bound students who need a strong background before entering the university. Davis stressed again that students are capable of learning higher level mathematics and more mathematics than we generally expect of them, and he suggested we challenge our students at all grade levels by introducing topics traditionally taught at higher levels to children at earlier ages.

Nevertheless, "pushed down" mathematics as promoted by the "new math" materials did establish a foundation in our current curriculum. For example, solid and plane geometry are now taught in one course in 9th or 10th grade, and trigonometry has been placed with advanced algebra in 11th grade. This has allowed more advanced topics such as calculus to be taught during 12th grade. "It was not many years ago that the undergraduate mathematics program in an American college or university did not go beyond the calculus"[25,p.475]. Now, we see that many high school students have completed calculus before entering college. Some final year secondary students take a class called pre-calculus. This course is a direct descendent of the "new math" course covering topics such as functions and analysis[26]. Therefore, the movement begun as part of the "new mathematics" to push down higher level mathematics has been achieved, and this new format of mathematics structure in courses is widely taught today.

Besides the "pushed down" curriculum, the *Standards* also encourages topics that were first suggested during the "new math" movement. These include the desire for students to learn probability and statistics.   Furthermore, the number line was emphasized by SMSG, and it is prevalent in almost every mathematics classroom today[27].   Topics such as measurement, number, and estimation are also encouraged throughout the *Standards*.

In order to help teachers implement the ideas in the *Standards* more easily, NCTM published an Addenda Series and a Navigation Series that was developed to provide instructional ideas and materials for use in the classroom.   In a practical way, these series address teaching method, content, and assessment.   The materials were field tested to make them more "teacher friendly".   The writing teams for this series included teachers, supervisors, and university mathematics educators.   The Addenda Series offered one book per grade level for grades K-6 as well as fifteen subject area books for grade bands K-4, 5-8, and 9-12 concerning standards that may be harder to implement without supplements.   The Navigation Series is currently being developed with books based in topics for each of the four grade bands. In general, the process strands are woven throughout the examples and books[28].

Another similarity with the "new math" era is the inclusion of math structure which emphasized some of the abstract ideas that had been so popular in the 1960s.   The structure of mathematics stresses two interrelated ideas: one is the mathematical properties of manipulating expressions and the other is that the student can explain and understand the mathematics involved in using these properties. The *Standards* also promotes mathematical structure by expecting students to compute and perform mathematical operations with an understanding of what they are doing.

While the original "new math" materials dwelled on abstraction, the "new math" materials also tried to incorporate applied or "real world" problems.   The "new math" movement is often credited with offering many more problems that were related to "real life" than the traditional curriculum included[29].   The *Standards* also encourages teachers to relate problems to students' experiences and future experiences when possible.   For example, through the connections and communications standards, students should be able to build on previous knowledge while understanding "new mathematics" by making these connections and explaining them.

Another similarity is the emphasis on language.   The "new math" movement stressed the importance of precise language in its

materials.   One of the process standards of the *Standards* is the
importance of mathematics as communication.   Students must be able
to communicate their ideas effectively to others.   Providing an answer
is not enough—students must be able to explain their answer and why
they chose particular techniques to reach an answer or conclusion.
Furthermore, the language of mathematics still strives to be
unambiguous.   For example, teachers are encouraging students to use
terms such as "regroup" and "rename" instead of "borrow" and "carry."
Therefore, effective and precise language is necessary in
communication.

Both the "new math" movement and the *Standards* movement
shifted emphasis away from computation as the goal to understanding
of mathematics as the goal.   In doing so, as with the "new math"
movement, the *Standards* has its share of critics who do not feel that
NCTM emphasizes basic skills (computation and calculation) enough.
In fact, the current "math wars" often revolve around the issue of
computation in the curriculum because traditional educators have felt
that NCTM has tried to eliminate computation[30].   To help solve this
problem, the 2000 *Standards* specifically stresses that basic skills
include not only the ability to compute but also the ability to problem
solve.   Furthermore, they stress that computational fluency should
develop along with the understanding of mathematical operations.

Assessment Similarities and Differences

Finally, assessment of the "new math" movement and the
Standards movement will be discussed (see Table 5).   Many of the
reports of the 1980s showed that American children were not
performing as well in mathematics in relation to students from other
countries.   *The Second International Mathematics Study, A Nation at
Risk,* and *Educating Americans for the 21st Century* all portrayed the
mathematical abilities of American children poorly[31].   Most of these
international comparisons were based on standardized testing for at
least one means of comparison.   Possibly as a result of these tests and
reports, NCTM has advocated for alternative forms of assessment.

NCTM published *Assessment Standards for School
Mathematics* in 1995 in which they emphasized the need for multiple
sources of assessment to determine what students were learning.
NCTM stressed that tests alone were an inadequate way of assessing a

Table 5.
Assessment of the "New Math" Movement Compared to Assessment of the *Standards* Movement

"New Math"	*Standards*
The "new math" programs realized that students needed to be assessed differently; however, they rarely allowed the opportunity and instead focused on traditional testing to prove that their materials were not detrimental to students.	With the promotion of non-traditional teaching techniques such as discovery teaching rather than direct instruction, the *Standards* have called for different kinds of assessments.
Most testing by the projects during the "new math" era showed that students involved in the programs performed as well as students in traditional programs on traditional tests.	The *Standards* promotes that students are performing equally as well in mathematics as before, but that various kind of assessments are necessary in order to show adequately what students are learning.
The "new math" era was dominated by standardized testing and a lack of multiple means of assessment.	The *Standards* promotes multiple means of assessment such as performance based, interviews, spiraling, etc. instead of relying on one type of test.
Similarities between "new math" assessment and the *Standards* are not as pronounced because of the lack of attention paid towards assessment during the "new math" era.	The *Standards* have possibly learned from the "new math" movement by stressing the importance of various type of assessments to adequately measure what is being asked of students.

student's learning process, and other evidence was crucial. Other types of assessment such as observations, discussions, and examination of homework is necessary along with traditional tests[32].

NCTM promotes that assessments must reflect what we are expecting students to learn; therefore, assessments must ask students to explain and show their reasoning rather than traditionally asking for answers testing memorization or computation[33]. Many states are altering their examinations to require students to explain answers—showing more understanding of concepts. NCTM strongly advocates that assessments should reflect what we are asking students to learn—a situation that most would agree never happened in testing the "new math" materials and curriculums. During the "new math" era, project leaders expected to test their materials by means of the traditional assessment, the standardized test. In that case, testing did not adequately show what the students had learned, and other means of assessment were necessary to show the success of the "new math" programs. Appropriate assessment was an issue concerning success of the "new math" and is equally important presently. Nevertheless, assessment issues are controversial and debated today.

From the above analysis for the categories of philosophy, teachers, pedagogy, assessment, and mathematics content, similarities and differences between the "new math" movement and the *Standards* movement are apparent. Similarities are more predominant than differences, yet both demonstrate a connection between the "new math" and the *Standards* movements.

Summary

While the "new math" movement had both positive and negative effects, there was overall a more positive than negative mathematical influence. However, educators must remember which were the positive developments of the "new mathematics" movement and should provide a guide for future reform rather than allowing educators to "repeat history." Much of mathematics reform appears to be based on a swinging pendulum with constructivism and behaviorism at opposite ends. The NACOME report alluded to the problems of dichotomies within the mathematics community. This report stated that many educators felt that they must be at one end or the other of the pendulum. For example, some believe that they must choose between

"the old or the new in mathematics"
"skills or concepts"
"the concrete or the abstract"
"structure or problem-solving"
"induction or deduction" [34, p.136]

Therefore, according to some educators, each of these lies at one end of a pendulum. In order for educators to truly learn from mathematics reform and guide future reform, they must be able to choose within the pendulum swing and not restrict themselves to the two extremes. A battle has continually raged as this pendulum swings between "drill and practice" and "pure discovery learning". However, as mathematics education changes and improves, we must be able to learn from reform rather than immediately abandon it. "As we work to improve school mathematics, we should retain good ideas from previous curricula and not let them be swept aside by the pendulum of reform"[35,p.45].

While many educators and parents believe that the "new mathematics" disappeared years ago, it is, in fact, still very present in our standard curriculum and the structure of mathematics classes taught in schools. Furthermore, mathematics reform is as crucial today as it was in the 1960s but is continuing with more caution and educational expertise than before. We will and should continue to improve mathematics with continual mathematical reform. The following quote from 1965 could easily apply today:

> They brought with them many ideas, values, prejudices, and beliefs that were often quite at variance with usual pre-college notions. The impact of this confrontation (or collision) can still be felt in tremors and vibrations that portend a *very* different future for American education, if we can understand what is *really* happening, and if we have the wisdom and courage to build upon it and to grow with it[36,p.11].

Conclusion

Chapter Six provides a summary of the major findings from this historical research.

Summary

The author began by analyzing historical time periods shortly before the "new math" movement, during the "new math" movement, and shortly after the "new math" movement. As part of placing the "new math" movement in context, the author explained the general curriculum and traditional mathematics curriculum in the United States as well as the teaching standards and pedagogical standards before the "new math" movement.

Next, the author discussed the many "new math" projects of the 1950s and 1960s whose conclusions can be formed into the explanation of the "new math" movement as a whole. By explaining the many projects and curriculum groups contributing to this movement, the author was able to draw general conclusions in order to compare "new math" findings with the current reform movement. A strong history of the many "new math" projects needed to be completed in order to demonstrate the results of the "new math" projects and materials and provide explanations surrounding the "new math" era.

Next the author compared these two movements; showing the many similarities and some differences between the movements. The author showed the comparisons by referencing the NCTM *Standards* publications. Much of the analysis surrounding the "new math" movement was made from studying historical references—many of which were original documents from the 1950s and 1960s projects. This thorough historical review provided a solid history of the "new math" movement based on the "new math" projects. The most important reason to provide a historical review of the "new math" movement was to finally allow a fair comparison between it and another reform movement. Without a solid history of the "new math" movement, previous comparisons are often grounded in opinion rather than fact. By analyzing the two reform movements, specific conclusions were made.

Conclusions

Did the "new mathematics" movement fail? Actually, many of the new topics of the "modern math" era are present today[1]. The material that was "pushed down" in the 1960s as well as the traditional material taught from a new point of view in the 1960s is present in our current programs. Several positive outcomes of the "new math" include stronger qualifications for mathematics teachers of all levels; an increase in the number of mathematics educational specialists; and the many curricular reforms being promoted by various other professions. The "new math" era was a time when many educators focused more on content and pedagogy than before. "The strength of mathematics education in the U.S. today is due in no small part to the recruitment into the community during the new math era of some of the most talented people of that generation"[2, p2. of Lesson 5]. Furthermore, the "new math" movement brought mathematicians or university professors into the schools—a relationship that should be appreciated and utilized[3]. Likewise, currently Arts and Sciences faculty members are encouraged to collaborate with the teacher preparation programs.

Therefore, while the general public believes that "new math" failed, there is evidence that it did not—a look at the *Standards* and modern reform shows many aspects are very similar. In fact, those adverse to NCTM and its *Standards* have labeled this current reform movement the "new-new math"[4].

"The New Math and the *Standards* are related to each other. Many ideas—like thinking and problem solving—were emphasized both then and now"[5, p. 197]. The following information relates to general aims of mathematics instruction as expressed by many of the "new mathematics" projects[6]. The student should 1) find joy in mathematics; 2) understand mathematics and apply it to other situations; 3) develop assurance in his or her own work; and 4) be prepared to apply mathematics practically. Furthermore, most of the mathematics projects of the 1960s promoted problem solving in the context of applied mathematical interconnections. Many of these aims are very similar to the goals expressed in the *Standards*—indicating that the main goals of mathematics reform are similar to both the present and the "new mathematics" era: to develop students to their highest mathematical abilities through the promotion of problem solving so that students understand mathematics.

The primary question at the beginning of this book was, "What are the connections between the "new mathematics" movement and the *Standards*?" The "new math" movement and the current reform established by the National Council of Teachers of Mathematics and its *Standards* have similar themes. As explained in the previous chapter, the "new math" philosophies, mathematics content, teacher changes, pedagogy, and assessments have been similar to the *Standards* movement or affected the changes in the *Standards* movement.

Both movements had an underlying philosophy that the goal of mathematics education should be that students truly understand mathematics, and that they learn to communicate and problem solve based on their discoveries in mathematics. Prior to the "new math," arithmetic by rote learning was the primary focus in mathematics classrooms. While the "new math" movement began with focus on higher ability students, it evolved toward focus on mathematics for all—a foundation of NCTM. Both movements called for a "pushed down" curriculum in mathematics offering students a more challenging curriculum—often in an integrated form so that students can see the connections among mathematics topics and between mathematics and other areas such as science. Both movements called for an increase in teacher knowledge of mathematics for in-service and pre-service teacher education programs. Both movements promoted the pedagogical philosophy of constructivism that uses techniques such as discovery learning, cooperative learning, and hands-on development from concrete thought to abstract thought to provide a student-centered classroom. Both movements saw the need for multiple assessments in order to adequately and fairly assess students' learning from a constructivist view rather than the traditional view; however, the *Standards* has been able to promote this idea further.

Therefore, "many observers today see strong similarities between the current movement for academic excellence and the post-*Sputnik* reforms"[7,p.83]. Some differences exist as well. For example, the "new math" movement was led by mathematicians rather than mathematics educators. The "new math" movement began as a curriculum for more advanced students. However, the largest difference between the two movements has been assessment. The "new math" projects continually tried to prove that their new curriculum would not negatively affect student performance on standardized test scores. However, NCTM promotes that other kinds of assessments must be completed in order to accurately test what children are learning because what and how they are learning is different. Besides these few differences, this historical analysis of the "new mathematics" era did show the influences on modern day mathematical reform.

We are presently deciding which direction the *Standards* will ultimately lead mathematics education. The "new math" movement, though popular, did not change a large majority of schools in the United States. The sudden backlash and over-reaction to the "new math" movement possibly did not allow enough time to fully develop. When analyzing the influence the *Standards* has had, many authors of textbooks and journals indicate that it has had a large effect. The Third International Mathematics and Science Study found that a large majority of teachers were aware of the current reform and many felt they were using those ideas in their classroom[8]. However, when McLeod analyzed schools and interviews with state supervisors of mathematics, it seemed that there were still many classrooms where the *Standards* has not had much success[9]. Therefore, while the *Standards* have impacted mathematics education, more time and emphasis may be necessary to see change.

"The improvement of the mathematics curriculum, like the growth of mathematics itself, is a never-ending process"[10,p.60]. Curriculum development is exactly what it says: continual development of the curriculum. There is need for continued mathematics reform. Furthermore, we must understand that "basic skills" does not mean only arithmetic and computation. Basic skills, as defined by NCTM in the *Principles and Standards* are the ability to reason, problem solve, make connections, communicate, and understand mathematics while using the tools of computational fluency. Therefore, basic skills means increasing a student's ability to face a problem and begin to solve the problem because of past knowledge and experience in mathematics reasoning and problem solving with competence in computation. Because the current mathematics reform (and the "new math" reform) stresses more creativity, the traditional elements and abilities to compute are not absent. Educators realize the importance of all skills in mathematics for all children, and few reforms advocate for the elimination of computation.

In order to avoid repeating mistakes, it is the author's recommendation that the *Standards* persevere in order to allow enough time for mathematics education change. We are at a stage in mathematics reform where some have advocated again for a "back to basics" movement[11]. Rather than hastily leaving the *Standards*, Bosse has suggested to allow educators more time rather than repeating the mistake of abandoning the "new math" before adequate time was allowed for obvious improvement in mathematics and problem solving skills.

Another area we should focus is anxiety of children in learning mathematics. While much emphasis has been placed on content and pedagogy, math anxiety of students should be addressed in order to create more positive mathematics attitudes and less fear for longer periods of time. Therefore, educators promoting the *Standards* must address anxieties of students who are discovering and actively participating in their learning.

Parents, teachers, educators, etc. must believe themselves that mathematics can be fun and challenging and taught as a creative subject. Furthermore, teachers and parents must communicate more effectively—particularly so that parents can understand that the mathematics their children may be studying will be different than the mathematics they studied. "We laugh about parents demanding that their children suffer mathematics in the same way that they did. But we should be hearing the genuine desire to help their children that is underneath their words"[12,p.9]. Many parents were taught computation and arithmetic as the main goals of mathematics; therefore, they feel most comfortable helping their children with similar problems. Today, NCTM offers assistance to parents through their web site (www.nct.org), and other programs such as "Figure This!" are available to help parents with working with their children on mathematics[13].

In order for any reform to occur successfully, both teachers and students must be willing to accept and understand the changes.

Teachers must have ownership of the reform and be willing to strengthen their teaching techniques. Ideas such as guided discovery cannot be explained properly to a teacher who is reading it for the first time in a new mathematics lesson—teachers must be willing to learn new methods and techniques because they believe in the reform[14]. We still need to focus on the problem of creating highly capable and qualified teachers.

In conclusion, the "new math" movement and the *Standards* movement have similar philosophies and goals. Both movements have been dominated by enthusiasm for improvement in mathematics education. Unlike the "new math" movement, we should not abandon the current reform but rather, contribute our own knowledge as teachers, mathematics educators, and parents while encouraging children in this *Standards* curriculum.

Most importantly, let us as educators stop continually repeating history—swinging back and forth on the pendulum described previously. Let us continue to develop a solid mathematics framework, "taking the best from the past with the knowledge we have gained since then to educate those who will follow us"[15,p.181]. While mathematics education is complex and often confusing because of the many forces pulling on which direction reform should lead, we must remember that the focus of mathematics education must be on successful teaching of mathematics for children. The development of any mathematics curriculum reform is a matter of evolution. "As with all evolution, the end is not clearly in sight—evolution proceeds by continually modifying the existing situation, and the 'end result' is not determined in advance"[16,p.38].

Appendix A

A summary of the content in the 1989 *Standards*

Goals for students:
1. Learn to value mathematics
2. Become confident in their abilities in mathematics
3. Become problem solvers in mathematics
4. Learn to communicate mathematically
5. Learn to reason mathematically

The first four standards are the same for grades K–12:
1. Mathematical Problem Solving
2. Mathematical Communication
3. Mathematical Reasoning
4. Mathematical Connections

The remaining standards are different for grades K–12, and they are all mathematically content specific. These are:

K-4
- Estimation
- Number sense and numeration
- Concepts of whole number operations
- Whole number computation
- Geometry and spatial sense
- Measurement
- Statistics and probability
- Fractions and decimals
- Patterns and relationships

5-8
- Number and number relationships
- Number systems and number theory
- Computation and estimation

- Patterns and functions
- Algebra
- Statistics
- Probability
- Geometry
- Measurement

9-12
- Algebra
- Functions
- Geometry from a synthetic perspective
- Geometry from an algebraic perspective
- Trigonometry
- Statistics
- Probability
- Discrete mathematics
- Conceptual underpinnings of calculus
- Mathematical structure

Appendix B
A summary of the content in the 2000 *Standards*

Principles for School Mathematics:

1. Equity Principle (high expectations and strong support for all students)
2. Curriculum Principle (coherent and focused curriculum across all grades)
3. Teaching Principle (effective teachers who understand mathematics teaching and learning across all grades)
4. Learning Principle (understanding mathematics based on prior knowledge)
5. Assessment Principle (various assessments providing useful information to both students and teachers)
6. Technology Principle (appropriate technology to enhance student learning)

Standards for School Mathematics (Grades Pre-K through 12):

Content Standards	Process Standards
Number and Operations	Problem Solving
Algebra	Reasoning and Proof
Geometry	Communication
Measurement	Connections
Data Analysis and Probability	Representation

All Standards are encompassed for each of the four grade-level bands:

1. Pre-kindergarten through Grade 2
2. Grades 3–5
3. Grades 6–8
4. Grades 9–12

While these standards are the same in title for each grade level band, specific differences and similarities are discussed in detail for each grade level band.

Appendix C
A Glossary of Acronyms

AMS	American Mathematical Society
BCMI	Boston College Mathematics Institute
CCSM	Cambridge Conference on School Mathematics
CCTT	Cambridge Conference on Teacher Training
CEEB	College Entrance Examination Board
CEMREL	Central Midwestern Regional Educational Laboratory
CSMP	Comprehensive School Mathematics Project
CUPM	Committee on the Undergraduate Program in Mathematics
ESEA	Elementary and Secondary Education Act
ESI	Educational Services Incorporated
GCMP	Greater Cleveland Mathematics Project
MAA	Mathematical Association of America
MINNEMAST	Minnesota Mathematics and Science Teaching Project
NACOME	National Advisory Committee on Mathematics Education
NCTM	National Council of Teachers of Mathematics
NDEA	National Defense Education Act
NSF	National Science Foundation
SMP	School Mathematics Project
SMSG	School Mathematics Study Group
SSMCIS	Secondary School Mathematics Curriculum Improvement Study
UICSM	University of Illinois Committee on School Mathematics
UMMaP	University of Maryland Mathematics Project

Notes

Chapter I: Introduction

1.	Jack Silverberg, "Student's awed by big words and big ideas," *Winnipeg Free Press* (May 13, 1966) (University of Texas at Austin Archives, box 28): 6.

2.	NCTM, "The revolution in school mathematics," (Washington, D.C.: A report of Regional Orientation Conferences in Mathematics, 1961): 1-90.

3.	Elisabeth Barlage, "The new math. An historical account of the reform of mathematics instruction in the United States of America," (*ERIC Document Reproduction Service No. ED 224 703*). (1982).

4.	Morris Kline, *Why Johnny can't add: The failure of the new math* (New York: St. Martin's Press, 1973).

5.	Robert W. Hayden, "A historical view of the new mathematics." (*ERIC Document Reproduction Service No. ED 228 046*), (Paper presented at the American Educational Research Association Symposium, 1983).

6.	Phillip S. Jones, *A history of mathematics education in the United States and Canada*, vol. 32nd yearbook (Washington, D.C.: National Council of Teachers of Mathematics, 1970).

7.	Morris Kline, *Why Johnny can't add: The failure of the new math* (New York: St. Martin's Press, 1973).

8.	Robert W. Hayden, "A history of the "new math" movement in the United States" (Doctor of Philosophy, Iowa State University, 1981).

9.	Edwin E. Moise et al., *Five views of the "New Math"* (Washington, D.C.: Council for Basic Education, 1965). (Maryville University School of Education Archives).

10.	Phillip S. Jones, *A history of mathematics education in the United States and Canada*, vol. 32nd yearbook (Washington, D.C.: National Council of Teachers of Mathematics, 1970).

11.	Max A. Sobel and Evan M. Maletsky, *Teaching mathematics*, 3rd ed. (Boston: Allyn and Bacon, 1999).

12.	Arthur F. Coxford, "Mathematics curriculum reform," in *National Council of Teachers of Mathematics history volume: a recent history of mathematics education in the United States and Canada*, ed. George M.A. Stanic and Jeremy Kilpatrick (Reston, VA: National Council of Teacher of Mathematics, in press).

13.	Joseph G.R. Martinez and Nancy C. Martinez, "In defense of mathematics reform and the NCTM's Standards," *The Mathematics Teacher* 91, no. 9 (1998): 746-748.

14.   Robert W. Hayden, "A historical view of the new mathematics." (*ERIC Document Reproduction Service No. ED 228 046), (*Paper presented at the American Educational Research Association Symposium, 1983).

Chapter II: "New Math" History

1.   F.V. Pohle, "Mathematics and the high school curriculum," (Garden City, NY: Adelphi University, 1965), (University of Texas at Austin Archives, box 37).
2.   Robert W. Hayden, "A historical view of the new mathematics." (*ERIC Document Reproduction Service No. ED 228 046), (*Paper presented at the American Educational Research Association Symposium, 1983).
3.   Phillip S. Jones, *A history of mathematics education in the United States and Canada*, vol. 32nd yearbook (Washington, D.C.: National Council of Teachers of Mathematics, 1970).
4.   Robert W. Hayden, "A history of the "new math" movement in the United States" (Doctor of Philosophy, Iowa State University, 1981).
5.   Carl B. Allendoerfer, "The second revolution in mathematics," *The Mathematics Teacher* 58, no. 8 (1965): 690-695.
6.   David Burner, Elizabeth Fox-Genovese, and Virginia Bernhard, *A college history of the United States* (St. James, NY: Brandywine Press, 1991).
7.   Richard N. Current et al., *The essentials of American history since 1865*, 4th ed. (New York: Alfred A. Knopf, 1986).
8.   George Brown Tindall, *America: A narrative history* (New York: W.W. Norton & Co, 1984).
9.   David Burner, Elizabeth Fox-Genovese, and Virginia Bernhard, *A college history of the United States* (St. James, NY: Brandywine Press, 1991).
10.   Richard N. Current et al., *The essentials of American history since 1865*, 4th ed. (New York: Alfred A. Knopf, 1986).
11.   Robert W. Hayden, "A history of the "new math" movement in the United States" (Doctor of Philosophy, Iowa State University, 1981).
12.   Richard N. Current et al., *The essentials of American history since 1865*, 4th ed. (New York: Alfred A. Knopf, 1986).
13.   David Burner, Elizabeth Fox-Genovese, and Virginia Bernhard, *A college history of the United States* (St. James, NY: Brandywine Press, 1991).

14.   Hillier Krieghbaum and Hugh Rawson, *An investment in knowledge* (New York: New York University Press, 1969).

15.   William Wooton, *SMSG: The making of a curriculum* (New Haven: Yale University Press, 1965).

16.   George Brown Tindall, *America: A narrative history* (New York: W.W. Norton & Co, 1984).

17.   UICSM, File of general information articles, University of Illinois Archives (1950s-1970s).

18.   Hillier Krieghbaum and Hugh Rawson, *An investment in knowledge* (New York: New York University Press, 1969).

19.   Mary Margaret Grady Nee, "The development of secondary school mathematics education in the United States, 1950-1965: Origins of policies in historical perspective" (Doctor of Philosophy, Loyola University, 1990).

20.   Glenda Lappan, *Lessons from the Sputnik era in mathematics education* [online] (Symposium of "Reflecting on Sputnik," Oct. 4, 1997 [cited July 22, 2000]). Available www.nas.edu/sputnik/

21.   Hillier Krieghbaum and Hugh Rawson, *An investment in knowledge* (New York: New York University Press, 1969).

22.   Richard N. Current et al., *The essentials of American history since 1865*, 4th ed. (New York: Alfred A. Knopf, 1986).

23.   CEEB, "Program for college preparatory mathematics," in *Readings in the history of mathematics education*, ed. James K. Bidwell and Robert G. Clason (Washington, D.C.: National Council of Teachers of Mathematics, 1970): 664-706.

24.   Richard N. Current et al., *The essentials of American history since 1865*, 4th ed. (New York: Alfred A. Knopf, 1986).

25.   Allister McMullen, "The University of Illinois Committee on School Mathematics project," (Report submitted to the Australian Council for Educational Research, 1958), (Urbana, IL: UICSM. Record series 10/13/1, box 3, University of Illinois Archives).

26.   Richard N. Current et al., *The essentials of American history since 1865*, 4th ed. (New York: Alfred A. Knopf, 1986).

27.   George Brown Tindall, *America: A narrative history* (New York: W.W. Norton & Co, 1984).

28.   Decker Walker, *Fundamentals of curriculum* (Orlando: Harcourt Brace Jovanovich, 1990).

29.   Richard N. Current et al., *The essentials of American history since 1865*, 4th ed. (New York: Alfred A. Knopf, 1986).

30.   James Bryant Conant, *The American high school today* (New York: McGraw Hill, 1959).

31.   NCTM, "The revolution in school mathematics," (Washington, D.C.: A report of Regional Orientation Conferences in Mathematics, 1961): 1-90.

32.   Mary Margaret Grady Nee, "The development of secondary school mathematics education in the United States, 1950-1965: Origins of policies in historical perspective" (Doctor of Philosophy, Loyola University, 1990).

33.   NCTM, "The revolution in school mathematics," (Washington, D.C.: A report of Regional Orientation Conferences in Mathematics, 1961): 1-90.

34.   Myron F. Rosskopf, "Technical mathematics for grades 9, 10, and 11," *School Science and Mathematics* (Record Series 10/13/1, Box 3, University of Illinois Archives, 1954): 594-600.

35.   Phillip S. Jones, *A history of mathematics education in the United States and Canada*, vol. 32nd yearbook (Washington, D.C.: National Council of Teachers of Mathematics, 1970).

36.   Edgar S. Leach, "Programs for talented mathematics students," (Evanston, IL: Record Series 10/13/5, Box 11, University of Illinois Archives).

37.   William Wooton, *SMSG: The making of a curriculum* (New Haven: Yale University Press, 1965).

38.   Phillip S. Jones, *A history of mathematics education in the United States and Canada*, vol. 32nd yearbook (Washington, D.C.: National Council of Teachers of Mathematics, 1970).

39.   Saunders MacLane, "The impact of modern mathematics," *The Bulletin* 38 (May 1954): 66-70.

40.   Max Beberman, "An integrated high school mathematics program," Speech given at 33rd Annual meeting of the Illinois MAA, (Galesburg, IL: Record Series 10/13/1, Box 8, University of Illinois Archives, 1954).

41.   NCTM and MAA, "The place of mathematics in secondary education," in *Readings in the history of mathematics education*, ed. James K. Bidwell and Robert G. Clason (Washington, D.C.: NCTM, 1970): 586-617.

42.   William David Reeve, "General mathematics in the secondary school, I," *Mathematics Teacher* 47, no. 2 (1954): 73-80.

43.   Helen Rowan, "The wonderful world of why: Where children are expected to think like children," *Saturday Review* 40, (Record Series 10/21/20, University of Illinois Archives, Nov. 2, 1957): 42-45.

44.   NCTM, "Guidance pamphlet in mathematics for high school students: The final report of the Commission on Post-War Plans of the National Council of Teachers of Mathematics," (New York: Record Series 10/13/1, Box 3, University of Illinois Archives, 1947).

45.   Robert W. Hayden, "A history of the "new math" movement in the United States" (Doctor of Philosophy, Iowa State University, 1981).

46.   NCTM, "The second report of The Commission on post-war plans," in *Readings in the history of mathematics education*, ed. James K. Bidwell and Robert G. Clason (Washington, D.C.: NCTM, 1970): 618-654.

47.   Phillip S. Jones, *A history of mathematics education in the United States and Canada*, vol. 32nd yearbook (Washington, D.C.: National Council of Teachers of Mathematics, 1970).

48.   CEEB, "Program for college preparatory mathematics," in *Readings in the history of mathematics education*, ed. James K. Bidwell and Robert G. Clason (Washington, D.C.: National Council of Teachers of Mathematics, 1970): 664-706.

49.   NCTM, "The revolution in school mathematics," (Washington, D.C.: A report of Regional Orientation Conferences in Mathematics, 1961): 1-90.

50.   William Wooton, *SMSG: The making of a curriculum* (New Haven: Yale University Press, 1965).

51.   Edwin E. Moise et al., *Five views of the "New Math"* (Washington, D.C.: Council for Basic Education, 1965). (Maryville University School of Education Archives).

52.   Robert W. Hayden, "A history of the "new math" movement in the United States" (Doctor of Philosophy, Iowa State University, 1981).

53.   NACOME, *Overview and analysis of school mathematics, grades K-12*, ed. National Advisory Committee on Mathematical Education Conference Board of the Mathematical Sciences (Reston, VA: NCTM, 1975).

54.   William Wooton, *SMSG: The making of a curriculum* (New Haven: Yale University Press, 1965).

55.   NCTM, "Guidance pamphlet in mathematics for high school students: The final report of the Commission on Post-War Plans of the National Council of Teachers of Mathematics," (New York: Record Series 10/13/1, Box 3, University of Illinois Archives, 1947).

56.   NCTM, "The revolution in school mathematics," (Washington, D.C.: A report of Regional Orientation Conferences in Mathematics, 1961): 1-90.

57.   University of Illinois College of Engineering, "Johnny can't do math problems either," *Engineering Outlook* 1, no. 5 (1960) (Record Series 10/13/1, box 3, University of Illinois Archives): 1-4.

58.   Robert E. Slavin, *Educational Psychology: Theory and Practice*, 5th ed. (Boston: Allyn & Bacon, 1997).

59.   UICSM, "The University of Illinois School Mathematics Program," in *Readings in the history of mathematics education*, ed. James K. Bidwell and Robert G. Clason (Washington, D.C.: NCTM, 1970).

60.   David L. Roberts, "Historical reflections on the role of discovery learning during the "new math" era: Rhetoric and reality," (paper presented at the AMS/MAA conference, Washington, D.C., January 2000).

61.   Carleton A. Chapman et al., "Report of the University High School parent survey: Committee on science and mathematics," (Urbana, IL: 1959, Record Series 10/13/1, Box 3, University of Illinois Archives).

62.   NACOME, *Overview and analysis of school mathematics, grades K-12*, ed. National Advisory Committee on Mathematical Education Conference Board of the Mathematical Sciences (Reston, VA: NCTM, 1975).

63.   Geoffrey Howson, Christine Keitel, and Jeremy Kilpatrick, *Curriculum development in mathematics* (Cambridge: Cambridge University Press, 1981).

64.   Edwin E. Moise et al., *Five views of the "New Math"* (Washington, D.C.: Council for Basic Education, 1965). (Maryville University School of Education Archives).

65.   Jeremy Kilpatrick, "A history of research in mathematics education," in *Handbook of research on mathematics teaching and learning*, ed. Douglas A. Grouws (New York: Macmillian Publishing Co, 1992): 3-38.

66.   Doris Diamant Machtinger, "Experimental course report, Kindergarten , no.2," (Webster Groves, MO: The Madison Project, 1965).

67.   Nuffield Mathematics Project, *I do, and I understand* (New York: John Wiley & Sons, Inc., 1967).

68.   Phillip S. Jones, *A history of mathematics education in the United States and Canada*, vol. 32nd yearbook (Washington, D.C.: National Council of Teachers of Mathematics, 1970).

69.   Geoffrey Howson, Christine Keitel, and Jeremy Kilpatrick, *Curriculum development in mathematics* (Cambridge: Cambridge University Press, 1981).

70.   Evelyn Sharp, *A parent's guide to more new math: Slide rules and peanut butter* (New York: E.P. Dutton & Co, 1966).

71.   NCTM, "The revolution in school mathematics," (Washington, D.C.: A report of Regional Orientation Conferences in Mathematics, 1961): 1-90.

72.   SMSG, "Report of a conference on elementary school mathematics" (Feb 13-14, 1959). (Record Series 10/12/1, Box 13, University of Illinois Archives).

73.   NCTM, "The revolution in school mathematics," (Washington, D.C.: A report of Regional Orientation Conferences in Mathematics, 1961): 1-90.

Chapter III: "New Math Projects"

1.   Myron F. Rosskopf, ed., *The teaching of secondary school mathematics*, vol. 33 (Reston, VA: National Council of Teachers of Mathematics, 1970).

2.   Bob Moon, *The 'new maths' curriculum controversy: an international story* (London: The Falmer Press, 1986).

3.   J. Fang, *Numbers racket: The aftermath of the "new math"* (Port Washington, NY: Kennikat Press, 1968).

4.   Robert W. Hayden, "A history of the "new math" movement in the United States" (Doctor of Philosophy, Iowa State University, 1981).

5.   NCTM, *An analysis of new mathematics programs* (Washington, D.C.: National Council of Teachers of Mathematics, 1963).

6.   UICSSM, "A project for the improvement of instruction, organization, and understanding of secondary school mathematics," (Urbana, IL: Record Series 10/13/1, Box 8, University of Illinois Archives, Summer 1954).

7.   Robert W. Hayden, "A history of the "new math" movement in the United States" (Doctor of Philosophy, Iowa State University, 1981).

8.   American Association for the Advancement of Science, *Second report of the information clearinghouse on new science curricula* (College Park, MD: AAAS and the University of Maryland Science Teaching Center, 1964).

9.   Thomas Steven Dupre, "The University of Illinois Committee on School Mathematics and the "new mathematics" controversy" (Doctor of Philosophy, University of Illinois, 1986).

10.   William A. Ferguson et al., "Mathematics needs of prospective students in the College of Engineering at the University of Illinois," *University of Illinois Bulletin* 49, no. 18 (Record series 10/13/1, box 8, University of Illinois Archives, 1951): 1-18.

11.   UICSM, "The work of the University of Illinois Committee on School Mathematics," (Urbana, IL: Record Series 10/13/1, Box 5, University of Illinois Archives, 1956).

12.   William A. Ferguson et al., "Mathematics needs of prospective students in the College of Engineering at the University of Illinois," *University of Illinois Bulletin* 49, no. 18 (Record series 10/13/1, box 8, University of Illinois Archives, 1951): 1-18.

13.   UICSM, "Information for high school students concerning the mathematics requirements in the College of Engineering at the University of Illinois," (Urbana, IL: Record Series 10/13/1, Box 5, University of Illinois Archives).

14.   Max Beberman, *An emerging program of secondary school mathematics* (Cambridge: Harvard University Press, 1958).

15.   Joe Wright, "Mathematics," *University of Illinois News*, March 3, 1955, (Record Series 10/13/1, Box 3, University of Illinois Archives).

16.   UICSM, "History of project," (Urbana, IL: Record Series 10/13/1, Box 8, University of Illinois Archives).

17.   Ross Taylor, "First course in algebra -- UICSM and SMSG: a comparison," *Mathematics Teacher* 55, no. 6 (1962): 478-481.

18.   Phillip S. Jones, *A history of mathematics education in the United States and Canada*, vol. 32nd yearbook (Washington, D.C.: National Council of Teachers of Mathematics, 1970).

19.   B.E. Meserve, "The mathematical needs of prospective engineering students," (Paper presented at the annual meeting of the American Society for Engineering Education), (Urbana, IL: Record Series 10/13/1, Box 8, University of Illinois Archives, 1954).

20.   UICSM, "Report to the Mathematical Association of America on the University of Illinois Committee on School Mathematics," (Washington, D.C.: 1958), (MAA Record Series, 10/13/1, Box 1, University of Illinois Archives).

21.   NCTM, "The revolution in school mathematics," (Washington, D.C.: A report of Regional Orientation Conferences in Mathematics, 1961): 1-90.

22.   Helen L. Garstens, M.L. Keedy, and John R. Mayor, "University of Maryland Mathematics Project," *Arithmetic Teacher* 7, no. 2 (1960): 61-70.

23.   Max Beberman, *An emerging program of secondary school mathematics* (Cambridge: Harvard University Press, 1958).

24.   W. Eugene Ferguson, "What should the science and mathematics curriculum be for the secondary schools of tomorrow?," (The Annual School Board Conference). (Record Series, 10/13/1, Box 13, University of Illinois Archives).

25.   UICSM, File of general information articles, University of Illinois Archives (1950s-1970s).

26.   Gertrude Hendrix, "The UICSM mathematics project films," (Urbana, IL: Record Series 10/13/10, Box 1, University of Illinois Archives).

27.   Phillip S. Jones, *A history of mathematics education in the United States and Canada*, vol. 32nd yearbook (Washington, D.C.: National Council of Teachers of Mathematics, 1970).

28.   UICSM, "The mathematics project at University High School," (Urbana, IL: Record Series 10/13/1, Box 8, University of Illinois Archives).

29.   Phillip S. Jones, *A history of mathematics education in the United States and Canada*, vol. 32nd yearbook (Washington, D.C.: National Council of Teachers of Mathematics, 1970).

30.   Kenneth E. Brown and Theodore L. Abell, *Analysis of research in the teaching of mathematics* (Washington, D.C.: U.S. Government Printing Office, 1965).

31.   UICSM, "Press release, 1961-1962" (Urbana, IL: Record Series 10/13/10, Box 1, University of Illinois Archives, 1961).

32.   UICSM, "Report to the Mathematical Association of America on the University of Illinois Committee on School Mathematics," (Washington, D.C.: 1958), (MAA Record Series, 10/13/1, Box 1, University of Illinois Archives).

33.   Phillip S. Jones, *A history of mathematics education in the United States and Canada*, vol. 32nd yearbook (Washington, D.C.: National Council of Teachers of Mathematics, 1970).

34.   Max Beberman, "Statement of the problem" (paper presented at the UICSM conference on the role of applications in a secondary school mathematics curriculum, Monticello, IL, Feb 14-19 1963), (Record Series 10/12/1, Box 21, University of Illinois Archives).

35.   Scott Foresman and Company, *Studies in mathematics education: A brief survey of improvement programs for school mathematics* (Chicago: Scott Foresman, 1960). (University of Maryville School of Education Archives).

36.   David A. Page, "Proposal to the Carnegie Corporation for the University of Illinois Arithmetic Project," (Urbana, IL: Record Series 10/12/1, Box 10, University of Illinois Archives, 1958).

37.   University of Illinois Arithmetic Project, "General information: The University of Illinois Arithmetic Project," (Urbana, IL: Record Series 10/12/7, Box 1, University of Illinois Archives, January, 1967).

38.   Scott Foresman and Company, *Studies in mathematics education: A brief survey of improvement programs for school mathematics* (Chicago: Scott Foresman, 1960). (University of Maryville School of Education Archives).

39.   University of Illinois Arithmetic Project, "General information: The University of Illinois Arithmetic Project," (Urbana, IL: Record Series 10/12/7, Box 1, University of Illinois Archives, January, 1967).

40.   Robert W. Hayden, "A history of the "new math" movement in the United States" (Doctor of Philosophy, Iowa State University, 1981).

41.   Helen L. Garstens, M.L. Keedy, and John R. Mayor, "University of Maryland Mathematics Project," *Arithmetic Teacher* 7, no. 2 (1960): 61-70.

42.   American Association for the Advancement of Science, *Second report of the information clearinghouse on new science curricula* (College Park, MD: AAAS and the University of Maryland Science Teaching Center, 1964).

43.   Scott Foresman and Company, *Studies in mathematics education: A brief survey of improvement programs for school mathematics* (Chicago: Scott Foresman, 1960). (University of Maryville School of Education Archives).

44.   Elisabeth Barlage, "The new math. An historical account of the reform of mathematics instruction in the United States of America," (*ERIC Document Reproduction Service No. ED 224 703*). (1982).

45.   Robert W. Hayden, "A history of the "new math" movement in the United States" (Doctor of Philosophy, Iowa State University, 1981).

46.   NCTM, *An analysis of new mathematics programs* (Washington, D.C.: National Council of Teachers of Mathematics, 1963).

47.   Helen L. Garstens, M.L. Keedy, and John R. Mayor, "University of Maryland Mathematics Project," *Arithmetic Teacher* 7, no. 2 (1960): 61-70.

48.   Richard M. Good and Helen M. Wolfle, "Evaluation of the University of Maryland eighth grade course in modern mathematics," (College Park, MD: Record Series 10/12/1, Box 1, University of Illinois Archives).

49.   UMMaP, "Progress report number 7," (College Park, MD: Record Series 10/12/1, Box 1, University of Illinois Archives, 1960).

50.   NCTM, *An analysis of new mathematics programs* (Washington, D.C.: National Council of Teachers of Mathematics, 1963).

51.   Richard M. Good and Helen M. Wolfle, "Evaluation of the University of Maryland eighth grade course in modern mathematics," (College Park, MD: Record Series 10/12/1, Box 1, University of Illinois Archives).

52.   Kenneth E. Brown and Theodore L. Abell, *Analysis of research in the teaching of mathematics* (Washington, D.C.: U.S. Government Printing Office, 1965).

53.   Phillip S. Jones, *A history of mathematics education in the United States and Canada*, vol. 32nd yearbook (Washington, D.C.: National Council of Teachers of Mathematics, 1970).

54.   William Wooton, *SMSG: The making of a curriculum* (New Haven: Yale University Press, 1965).

55.   Mary Margaret Grady Nee, "The development of secondary school mathematics education in the United States, 1950-1965: Origins of policies in historical perspective" (Doctor of Philosophy, Loyola University, 1990).

56.   Albert E. Meder, "Mathematics for today," (Record Series 10/13/1, Box 2, University of Illinois Archives): 1-4.

57.   Scott Foresman and Company, *Studies in mathematics education: A brief survey of improvement programs for school mathematics* (Chicago: Scott Foresman, 1960). (University of Maryville School of Education Archives).

58.   CEEB, "Program for college preparatory mathematics," in *Readings in the history of mathematics education*, ed. James K. Bidwell and Robert G. Clason (Washington, D.C.: National Council of Teachers of Mathematics, 1970): 664-706.

59.   CEEB, "A summary of the report of the commission on mathematics," (New York: Record Series 10/12/1, Box 2, University of Illinois Archives, 1959).

60.   CEEB, "Commission on mathematics," (New York: Record Series 10/13/1, Box 9, University of Illinois Archives, 1957).

61.   CEEB, "Objectives of the Commission on Mathematics of the College Entrance Examination Board," (New York: Record Series 10/13/1, Box 9, University of Illinois Archives, 1956).

62.   Robert W. Hayden, "A history of the "new math" movement in the United States" (Doctor of Philosophy, Iowa State University, 1981).

63.   Robert W. Hayden, "A historical view of the new mathematics." *(ERIC Document Reproduction Service No. ED 228 046),* (Paper presented at the American Educational Research Association Symposium, 1983).

64.   SMSG, "Information memorandum," (University of Texas at Austin Archives, 1963).

65.   SMSG, "Report of a conference on elementary school mathematics" (Feb 13-14, 1959). (Record Series 10/12/1, Box 13, University of Illinois Archives).

66.   Robert W. Hayden, "A history of the "new math" movement in the United States" (Doctor of Philosophy, Iowa State University, 1981).

67.   NCTM, *An analysis of new mathematics programs* (Washington, D.C.: National Council of Teachers of Mathematics, 1963).

68.   E.G. Begle, "The School Mathematics Study Group," (University of Texas at Austin Archives, Drawer 2:3).

69.   William Wooton, *SMSG: The making of a curriculum* (New Haven: Yale University Press, 1965).

70.   Carl B. Allendoerfer, "The second revolution in mathematics," *The Mathematics Teacher* 58, no. 8 (1965): 690-695.

71.   E.G. Begle, "The School Mathematics Study Group," (University of Texas at Austin Archives, Drawer 2:3).

72.   Phillip S. Jones, *A history of mathematics education in the United States and Canada*, vol. 32nd yearbook (Washington, D.C.: National Council of Teachers of Mathematics, 1970).

73.   Robert W. Hayden, "A history of the "new math" movement in the United States" (Doctor of Philosophy, Iowa State University, 1981).

74.   William Wooton, *SMSG: The making of a curriculum* (New Haven: Yale University Press, 1965).

75.   L.J. Paige, "SMSG," (The University of Texas at Austin Archives).

76.   William Wooton, *SMSG: The making of a curriculum* (New Haven: Yale University Press, 1965).

77.   E.G. Begle, *Critical variables in mathematics education: findings from a survey of the empirical literature* (Washington, D.C.: National Council of Teachers of Mathematics and Mathematics Association of America, 1979).

78.   Robert W. Hayden, "A history of the "new math" movement in the United States" (Doctor of Philosophy, Iowa State University, 1981).

79.   William Wooton, *SMSG: The making of a curriculum* (New Haven: Yale University Press, 1965).

80.   Lawrence Shirley, "Reviewing a century of mathematics education: Ready for the future," (paper presented at the National Council of Teachers of Mathematics Annual Meeting, Chicago, April 14, 2000).

81.   Sherman K. Stein, *Strength in numbers: Discovering the joy and power of mathematics in everyday life* (New York: John Wiley, 1996).

82.   Phillip S. Jones, *A history of mathematics education in the United States and Canada*, vol. 32nd yearbook (Washington, D.C.: National Council of Teachers of Mathematics, 1970).

83.   SMSG, "Committee on Latin American mathematics education," (Record Series 10/12/1, Box 13, University of Illinois Archives).

84.   Geoffrey Howson, Christine Keitel, and Jeremy Kilpatrick, *Curriculum development in mathematics* (Cambridge: Cambridge University Press, 1981).

85.   Mary Margaret Grady Nee, "The development of secondary school mathematics education in the United States, 1950-1965: Origins of policies in historical perspective" (Doctor of Philosophy, Loyola University, 1990).

86.   NCTM, *An analysis of new mathematics programs* (Washington, D.C.: National Council of Teachers of Mathematics, 1963).

87.   Noel Wical, "Jet age math will help 25,000 pupils here in fall," (*Cleveland State University Archives*, July 20, 1960).

88.   NCTM, *An analysis of new mathematics programs* (Washington, D.C.: National Council of Teachers of Mathematics, 1963).

89.   Robert B. Davis, "Final report," (D-233 OE-6-10-183). Washington D.C.:   U.S. Department of Health, Education, and Welfare. (Webster Groves, MO: Webster University Archives, 1967), 1-123.

90.   Caralee S. Stanard, "Webster College Public information announcements for the week of July 24," (Webster, MO: Webster College, Record Series 5/25/2, Webster University Archives, 1967).

91.   Webster College, "Dr. R. Davis joins WC faculty in fall," *The Web*, 1, (March 24, 1961) (Record series 5/25/2, Webster University Archives).

92.   Robert B. Davis, "A modern mathematics program as it pertains to the interrelationship of mathematical content, teaching methods and classroom atmosphere," (Webster Groves, MO: Webster University Archives, 1965), 1-112.

93.   Robert B. Davis, *Discovery in mathematics: A text for teachers* (Palo Alto: Addison-Wesley, 1964).

94.   Robert B. Davis, "A modern mathematics program as it pertains to the interrelationship of mathematical content, teaching methods and classroom atmosphere," (Webster Groves, MO: Webster University Archives, 1965), 1-112.

95.   Scott Foresman and Company, *Studies in mathematics education: A brief survey of improvement programs for school mathematics* (Chicago: Scott Foresman, 1960). (University of Maryville School of Education Archives).

96.   Madison Project, "Syracuse University-Webster College Madison Project progress report," (Webster Groves, MO, Record Series 12/12/1, Box 9, University of Illinois Archives, 1964).

97.   Madison Project, "Status report, Madison Project "implementation" program," (Webster Groves, MO, Record Series 10/13/1, Box 1, University of Illinois Archives, 1962).

98.   Doris Grundy, "Discovery technique is successfully applied to teaching of mathematics in Weston -- with amazing results," *Mark* (Record Series 5/25/5, Webster University Archives, July 1, 1961): 5-6.

99.   Robert B. Davis, "A modern mathematics program as it pertains to the interrelationship of mathematical content, teaching methods and classroom atmosphere," (Webster Groves, MO: Record Series 10/13/1, Box 1, University of Illinois Archives, 1962), 1-39.

100.   Robert B. Davis, *Explorations in mathematics: A text for teachers* (Palo Alto: Addison-Wesley, 1967).

101.   Robert B. Davis, "A brief introduction to materials and activities," (St. Louis, MO: The Madison Project, 1964).

102.   Robert B. Davis, "The evolution of school mathematics," *Journal of Research in Science Teaching* 1 (Record series 10/12/1, box 9, University of Illinois Archives, 1963): 260-264.

103.   Robert B. Davis, "A brief introduction to materials and activities," (St. Louis, MO: The Madison Project, 1964).

104.   Elisabeth Barlage, "The new math. An historical account of the reform of mathematics instruction in the United States of America," *(ERIC Document Reproduction Service No. ED 224 703).* (1982).

105.   Robert B. Davis, "A modern mathematics program as it pertains to the interrelationship of mathematical content, teaching methods and classroom atmosphere," (Webster Groves, MO: Webster University Archives, 1965), 1-112.

106.   Robert B. Davis, "A brief introduction to materials and activities," (St. Louis, MO: The Madison Project, 1964).

107.   Webster College, "Creativity marks traveling mathematician," *The Web* 3,)October 27, 1961) (Record series 5/25/2, Webster University Archives).

108.   Madison Project, "Syracuse University-Webster College Madison Project progress report," (Webster Groves, MO, Record Series 12/12/1, Box 9, University of Illinois Archives, 1964).

109.   Robert B. Davis, *Explorations in mathematics: A text for teachers* (Palo Alto: Addison-Wesley, 1967).

110.   Robert B. Davis, "Final report," (D-233 OE-6-10-183). Washington D.C.: U.S. Department of Health, Education, and Welfare. (Webster Groves, MO: Webster University Archives, 1967), 1-123.

111.   Robert B. Davis, *Discovery in mathematics: A text for teachers* (Palo Alto: Addison-Wesley, 1964).

112.   CEMREL, "Comprehensive school mathematics program: basic program plan," (St. Ann, MO: CEMREL, 1971).

113.   Phillip S. Jones, *A history of mathematics education in the United States and Canada*, vol. 32nd yearbook (Washington, D.C.: National Council of Teachers of Mathematics, 1970).

114.   Burt Kaufman, "Background information concerning CEMREL -- CSMP," (Carbondale, IL: Record Series 10/12/1, Box 3, University of Illinois Archives).

115.   Burt Kaufman, "The Comprehensive School Mathematics Program -- Its past, present, and future," (Carbondale, IL: CEMREL, 1967), (Maryville University School of Education Archives).

116.   Burt Kaufman, "The CSMP approach to a content oriented highly individualized mathematics education," (Carbondale, IL: CEMREL-CSMP, 1968), (Maryville University School of Education Archives).

117.   CSMP, "Comprehensive School Mathematics Program," (St. Louis, MO: CSMP Archives).

118.   CEMREL, "Comprehensive school mathematics program: basic program plan," (St. Ann, MO: CEMREL, 1971).

119.   Burt Kaufman and Nichola Sterling, "The CSMP elementary school program," (St. Louis, MO: CSMP Archives, 1978).

120.   CSMP, "Summary of comments of several SCMP staff and staff associates regarding questions posed for CEEB," (Record Series 10/12/1, Box 3, University of Illinois Archives, 1969-1970).

121.   Burt Kaufman, "The CSMP approach to a content oriented highly individualized mathematics education," (Carbondale, IL: CEMREL-CSMP, 1968), (Maryville University School of Education Archives).

122.   Phillip S. Jones, *A history of mathematics education in the United States and Canada*, vol. 32nd yearbook (Washington, D.C.: National Council of Teachers of Mathematics, 1970).

123.   Bryan Thwaites, *The School Mathematics Project: The first ten years* (Cambridge: Cambridge University Press, 1972).

124.   NCTM, "The revolution in school mathematics," (Washington, D.C.: A report of Regional Orientation Conferences in Mathematics, 1961): 1-90.

125.   NCTM, *An analysis of new mathematics programs* (Washington, D.C.: National Council of Teachers of Mathematics, 1963).

126.   Scott Foresman and Company, *Studies in mathematics education: A brief survey of improvement programs for school mathematics* (Chicago: Scott Foresman, 1960). (University of Maryville School of Education Archives).

127.   Geoffrey Howson, *A history of mathematics education in England* (Cambridge: Cambridge University Press, 1982).

128.   Bryan Thwaites, *The School Mathematics Project: The first ten years* (Cambridge: Cambridge University Press, 1972).

129.   Geoffrey Howson, Christine Keitel, and Jeremy Kilpatrick, *Curriculum development in mathematics* (Cambridge: Cambridge University Press, 1981).

130.   Bryan Thwaites, "The School Mathematics Project," (Southampton, England: 1963), (Record Series 10/13/1, Box 1, University of Illinois Archives).

131.   Bryan Thwaites, *The School Mathematics Project: The first ten years* (Cambridge: Cambridge University Press, 1972).

132.   Bryan Thwaites, "The School Mathematics Project," (Southampton, England: 1963), (Record Series 10/13/1, Box 1, University of Illinois Archives).

133.   NACOME, *Overview and analysis of school mathematics, grades K-12*, ed. National Advisory Committee on Mathematical Education Conference Board of the Mathematical Sciences (Reston, VA: NCTM, 1975).

134.   Robert B. Davis, "Final report," (D-233 OE-6-10-183). Washington D.C.:   U.S. Department of Health, Education, and Welfare. (Webster Groves, MO: Webster University Archives, 1967), 1-123.

135.   Nuffield Mathematics Project, *I do, and I understand* (New York: John Wiley & Sons, Inc., 1967).

136.   NCTM, *An analysis of new mathematics programs* (Washington, D.C.: National Council of Teachers of Mathematics, 1963).

137.   Elisabeth Barlage, "The new math. An historical account of the reform of mathematics instruction in the United States of America," (*ERIC Document Reproduction Service No. ED 224 703*). (1982).

138.   Phillip S. Jones, *A history of mathematics education in the United States and Canada*, vol. 32nd yearbook (Washington, D.C.: National Council of Teachers of Mathematics, 1970).

139.   Ball State University, "The Ball State Experimental Program in geometry and algebra," (Muncie, IN: Record Series 10/12/1, Box 2, University of Illinois Archives).

140.   Scott Foresman and Company, *Studies in mathematics education: A brief survey of improvement programs for school mathematics* (Chicago: Scott Foresman, 1960). (University of Maryville School of Education Archives).

141.   Ball State University, "Ball State," (Muncie, IN: Ball State Archives, 1965).

142.   Robert W. Hayden, "A history of the "new math" movement in the United States" (Doctor of Philosophy, Iowa State University, 1981).

143.   Scott Foresman and Company, *Studies in mathematics education: A brief survey of improvement programs for school mathematics* (Chicago: Scott Foresman, 1960). (University of Maryville School of Education Archives).

144.   B.E. Meserve, "The mathematical needs of prospective engineering students," (Paper presented at the annual meeting of the American Society for Engineering Education), (Urbana, IL: Record Series 10/13/1, Box 8, University of Illinois Archives, 1954).

145.   Scott Foresman and Company, *Studies in mathematics education: A brief survey of improvement programs for school mathematics* (Chicago: Scott Foresman, 1960). (University of Maryville School of Education Archives).

146.   Edwin E. Moise et al., *Five views of the "New Math"* (Washington, D.C.: Council for Basic Education, 1965). (Maryville University School of Education Archives).

147.   Fort Wayne Community Schools, "Author of newly adopted math program to meet with teachers," *Observer*, 9, no.3 (1970).

148.   Cambridge Conference on School Mathematics, "Goals for school mathematics," (paper presented at The Cambridge Conference on School Mathematics, Cambridge, Mass., 1963), (Record series 10/12/1, box 3, University of Illinois Archives).

149.   Educational   Services   Incorporated,   "Mathematics curriculum study: Cambridge Conference," (Educational Services Incorporated Quarterly Report, 1963), (Record series 10/12/1, box 4, University of Illinois Archives).

150.   Educational   Services   Incorporated,   *Goals for school mathematics* (Boston: Houghton Mifflin Company, 1963).

151.   Cambridge Conference on Teacher Training, "Goals for mathematical education of elementary school teachers," (Cambridge: Record Series 10/12/1, Box 4, University of Illinois Archives, January 9, 1967).

152.   Cambridge Conference on School Mathematics, "Draft of meeting summary," CCSM math-science meeting (Boston: Record Series 10/12/1, Box 4, University of Illinois Archives, 1965).

153.   Cambridge Conference on Teacher Training, "Goals for mathematical education of elementary school teachers," (Cambridge: Record Series 10/12/1, Box 4, University of Illinois Archives, January 9, 1967).

154.   Robert W. Hayden, "A history of the "new math" movement in the United States" (Doctor of Philosophy, Iowa State University, 1981).

155.   Elisabeth Barlage, "The new math. An historical account of the reform of mathematics instruction in the United States of America," (*ERIC Document Reproduction Service No. ED 224 703).* (1982).

156.   Scott Foresman and Company, *Studies in mathematics education: A brief survey of improvement programs for school mathematics* (Chicago: Scott Foresman, 1960). (University of Maryville School of Education Archives).

157.   Phillip S. Jones, *A history of mathematics education in the United States and Canada,* vol. 32nd yearbook (Washington, D.C.: National Council of Teachers of Mathematics, 1970).

158.   Scott Foresman and Company, *Studies in mathematics education: A brief survey of improvement programs for school mathematics* (Chicago: Scott Foresman, 1960). (University of Maryville School of Education Archives).

159.   NCTM,   "The   revolution   in   school   mathematics," (Washington, D.C.: A report of Regional Orientation Conferences in Mathematics, 1961): 1-90.

160.   W. Eugene Ferguson, "Implementing the new mathematics program in your school" (paper presented at the Regional Orientation Conferences in Mathematics, Fall 1960). (Record Series 10/13/1, box 2, University of Illinois Archives).

161.　　Katharine Doyle Rasch, "The evolution of selected "modern" mathematics content in elementary school mathematics textbooks 1963-1982" (Doctor of Philosophy, St. Louis University, 1983).

Chapter IV: Effects and Aftermath of the "New Math"

1.　　J. Fang, *Numbers racket: The aftermath of the "new math"* (Port Washington, NY: Kennikat Press, 1968).
2.　　Edwin E. Moise et al., *Five views of the "New Math"* (Washington, D.C.: Council for Basic Education, 1965). (Maryville University School of Education Archives).
3.　　Phillip S. Jones, *A history of mathematics education in the United States and Canada*, vol. 32nd yearbook (Washington, D.C.: National Council of Teachers of Mathematics, 1970).
4.　　Robert W. Hayden, "A history of the "new math" movement in the United States" (Doctor of Philosophy, Iowa State University, 1981).
5.　　Scott Foresman and Company, *Studies in mathematics education: A brief survey of improvement programs for school mathematics* (Chicago: Scott Foresman, 1960). (University of Maryville School of Education Archives).
6.　　Phillip S. Jones, *A history of mathematics education in the United States and Canada*, vol. 32nd yearbook (Washington, D.C.: National Council of Teachers of Mathematics, 1970).
7.　　Robert W. Hayden, "A history of the "new math" movement in the United States" (Doctor of Philosophy, Iowa State University, 1981).
8.　　Hillier Krieghbaum and Hugh Rawson, *An investment in knowledge* (New York: New York University Press, 1969).
9.　　Michael J. Bosse, "The NCTM Standards in light of the new math movement: A warning!," *Journal of Mathematical Behavior* 14, no. 2 (1995): 171-201.
10.　　Phillip S. Jones, *A history of mathematics education in the United States and Canada*, vol. 32nd yearbook (Washington, D.C.: National Council of Teachers of Mathematics, 1970).
11.　　CEEB, "The education of secondary school mathematics teachers," (New York: Record Series 10/13/1, Box 9, University of Illinois Archives, 1957).
12.　　Robert W. Hayden, "A history of the "new math" movement in the United States" (Doctor of Philosophy, Iowa State University, 1981).
13.　　Committee on the Undergraduate Program, "Conference on the Committee on the Undergraduate Program," *American Mathematical*

Monthly 66, no. 3 (1959), (Record Series 10/13/1, Box 1, University of Illinois Archives).

14.   Committee   on   the   Undergraduate   Program, "Recommendations   for   the   training   of   teachers   of   mathematics," (Mathematical Association of America, 1961), (Record Series 10/13/1, Box 12, University of Illinois Archives).

15.   Committee on the Undergraduate Program, "Pregraduate Training," (University of Texas at Austin Archives, Box 17, 1962).

16.   Committee   on   the   Undergraduate   Program, "Recommendations   for   the   training   of   teachers   of   mathematics," (Mathematical Association of America, 1961), (Record Series 10/13/1, Box 12, University of Illinois Archives).

17.   Robert J. Wisner, "CUPM -- It's activities and teacher training recommendations," *Committee on the Undergraduate Program of Mathematics Report, 1.* (Record Series 10/13/10, Box 4, University of Illinois Archives, Sept., 1961).

18.   Bert Y. Kersh, "Learning by discovery: What is learned?," (Speech presented at the meeting of the National Council of Teachers of Mathematics) (Eugene, OR: Record Series 10/13/1, Box 1, University of Illinois Archives, 1963).

19.   UICSM, "What is new in the "new mathematics"," *draft version for Compton's Encyclopedia Yearbook* (Record Series 10/13/1, Box 3, University of Illinois Archives, 1963).

20.   Myron F. Rosskopf, "Trends in content of high school mathematics in the United States," *Teachers College Record* 56, no. 3 (1954), (Record Series 10/13/1, box 3, University of Illinois Archives): 135-138.

21.   Edwin E. Moise et al., *Five views of the "New Math"* (Washington, D.C.: Council for Basic Education, 1965). (Maryville University School of Education Archives).

22.   Thomas   Steven   Dupre,   "The   University   of   Illinois Committee on School Mathematics and the "new mathematics" controversy" (Doctor of Philosophy, University of Illinois, 1986).

23.   Elisabeth Barlage, "The new math. An historical account of the reform of mathematics instruction in the United States of America," (*ERIC Document Reproduction Service No. ED 224 703).* (1982).

24.   Geoffrey Gould, "Beberman: Much of "new math" trivial, misdirected," *The News-Gazette,* April 11, 1966, (Champaign, IL: Record Series 10/21/20, University of Illinois Archives.

25.   UICSM, File of general information articles, University of Illinois Archives (1950s-1970s).

26.   Robert W. Hayden, "A history of the "new math" movement in the United States" (Doctor of Philosophy, Iowa State University, 1981).

27.   Morris Kline, "A criticism of S.M.S.G. and some positive suggestions," (University of Texas at Austin Archives).

28.   Stephen S. Willoughby, *Mathematics education for a changing world* (Alexandria, Virginia: ASCD, 1990).

29.   Morris Kline, *Why Johnny can't add: The failure of the new math* (New York: St. Martin's Press, 1973).

30.   Michael J. Bosse, "Reforming the NCTM Standards in light of historical perspective: Premature changes," *Journal of Mathematical Behavior* 17, no. 3 (1998): 317-327.

31.   Michael J. Bosse, "The NCTM Standards in light of the new math movement: A warning!," *Journal of Mathematical Behavior* 14, no. 2 (1995): 171-201.

32.   Lawrence Shirley, "Reviewing a century of mathematics education: Ready for the future," (paper presented at the National Council of Teachers of Mathematics Annual Meeting, Chicago, April 14, 2000).

33.   Morris Kline, "A criticism of S.M.S.G. and some positive suggestions," (University of Texas at Austin Archives).

34.   Phillip S. Jones, *A history of mathematics education in the United States and Canada*, vol. 32nd yearbook (Washington, D.C.: National Council of Teachers of Mathematics, 1970).

35.   Carl B. Allendoerfer, "The second revolution in mathematics," *The Mathematics Teacher* 58, no. 8 (1965): 690-695.

36.   Edwin E. Moise et al., *Five views of the "New Math"* (Washington, D.C.: Council for Basic Education, 1965). (Maryville University School of Education Archives).

37.   NCTM, "Proposals developed at meeting at Park Sheraton Hotel," (New York: Record Series 10/12/1, Box 13, University of Illinois Archives, 1964).

38.   Parents of UICSM students, "Letter to Max Beberman," (Urbana, IL: Record Series 10/13/1, Box 8, University of Illinois Archives, 1958).

39.   Morris Kline, *Why Johnny can't add: The failure of the new math* (New York: St. Martin's Press, 1973).

40.   Michael J. Bosse, "The NCTM Standards in light of the new math movement: A warning!," *Journal of Mathematical Behavior* 14, no. 2 (1995): 171-201.

41.   SMSG, "Report of a conference on elementary school mathematics" (Feb 13-14, 1959). (Record Series 10/12/1, Box 13, University of Illinois Archives).

42.   Michael J. Bosse, "Reforming the NCTM Standards in light of historical perspective: Premature changes," *Journal of Mathematical Behavior* 17, no. 3 (1998): 317-327.

43.   Jeremy Kilpatrick, *Five lessons from the new math era* [online] (Symposium of "Reflecting on Sputnik", Oct. 4, 1997 [cited July 22, 2000]). Available: www.nas.edu/sputnik/

44.   Jeremy Kilpatrick, "Confronting reform," *American Mathematical Monthly* 104 (1997): 955-962.

45.   Edwin E. Moise et al., *Five views of the "New Math"* (Washington, D.C.: Council for Basic Education, 1965). (Maryville University School of Education Archives).

46.   J.A. Easley, "Logic and heuristic in mathematics curriculum reform," Presented at the International Colloquium in the Philosophy of Science at Bedford College, London (July 11-17, 1965). (Record Series 10/13/1, Box 2, University of Illinois Archives).

47.   Carl B. Allendoerfer, "The second revolution in mathematics," *The Mathematics Teacher* 58, no. 8 (1965): 690-695.

48.   Robert W. Hayden, "A history of the "new math" movement in the United States" (Doctor of Philosophy, Iowa State University, 1981).

49.   Richard Nixon, *Education for the 1970's: Renewal and reform* (Washington, D.C.: U.S. Government Printing Office, 1970).

50.   Burt Kaufman and Nichola Sterling, "The CSMP elementary school program," (St. Louis, MO: CSMP Archives, 1978).

51.   NACOME, *Overview and analysis of school mathematics, grades K-12*, ed. National Advisory Committee on Mathematical Education Conference Board of the Mathematical Sciences (Reston, VA: NCTM, 1975).

52.   NCTM, *An agenda for action: Recommendations for school mathematics of the 1980s* (Reston, VA: National Council of Teachers of Mathematics , 1980).

53.   NCTM, *Priorities in school mathematics* (Reston, VA: National Council of Teachers of Mathematics, 1981).

54.   Douglas B. McLeod et al., "Setting the Standards," in *Bold ventures: Case studies of U.S. innovations in mathematics education*, ed. Senta Raizen and Edwards D. Britton (Dordrecht: Kluwer Academic Publishers, 1996): 376.

55.   National Research Council, *Everybody counts: A report to the nation on the future of mathematics education* (Washington, D.C.: National Academy Press, 1989).

56.   Douglas B. McLeod et al., "Setting the Standards," in *Bold ventures: Case studies of U.S. innovations in mathematics education*, ed. Senta Raizen and Edwards D. Britton (Dordrecht: Kluwer Academic Publishers, 1996): 376.

Chapter V: Connections with the *Standards*

1.   Douglas B. McLeod et al., "Setting the Standards," in *Bold ventures: Case studies of U.S. innovations in mathematics education*, ed. Senta Raizen and Edwards D. Britton (Dordrecht: Kluwer Academic Publishers, 1996): 376.

2.   NCTM, *Curriculum and evaluation standards for school mathematics* (Reston, VA: National Council of Teachers of Mathematics, 1989).

3.   Douglas B. McLeod et al., "Setting the Standards," in *Bold ventures: Case studies of U.S. innovations in mathematics education*, ed. Senta Raizen and Edwards D. Britton (Dordrecht: Kluwer Academic Publishers, 1996): 376.

4.   Joan Ferrini-Mundy, *"Principles and Standards for School Mathematics*: A guide for mathematicians," *Notices of the American Mathematical Society* 47, no. 5 (2000): 868-876.

5.   NCTM, *Principles and standards for school mathematics* (Reston, VA: National Council of Teachers of Mathematics, 2000).

6.   NCTM, "Answers to frequently asked questions about *Principles and Standards for School Mathematics*," *NCTM News Bulletin* 36, no. 9 (2000): 7-10.

7.   Susan Addington et al., "Four reactions to *Principles and Standards for School Mathematics*," *Notices of the American Mathematical Society* 4, no. 9 (2000): 1072-1079.

8.   Douglas B. McLeod et al., "Setting the Standards," in *Bold ventures: Case studies of U.S. innovations in mathematics education*, ed. Senta Raizen and Edwards D. Britton (Dordrecht: Kluwer Academic Publishers, 1996): 376.

9.   John A. Dossey et al., *The mathematics report card: Are we measuring up?*, ed. The National's Report Card, *National Assessment of Educational Progress* (Princeton: Educational Testing Service, 1988).

10.   NAEP (National Assessment of Educational Progress), *The nation's report card* [online] [cited October 1 2000]. Available http://nces.ed.gov/nationreportcard/site/home.asp

11.   Douglas B. McLeod et al., "Setting the Standards," in *Bold ventures: Case studies of U.S. innovations in mathematics education*, ed. Senta Raizen and Edwards D. Britton (Dordrecht: Kluwer Academic Publishers, 1996): 376.

12.   Christian R. Hirsch, ed., *Curriculum and evaluation standards for school mathematics: Addenda series grades 9 - 12* (Reston, VA: National Council of Teachers of Mathematics, 1991).

13.   Douglas B. McLeod et al., "Setting the Standards," in *Bold ventures: Case studies of U.S. innovations in mathematics education*, ed. Senta

Raizen and Edwards D. Britton (Dordrecht: Kluwer Academic Publishers, 1996): 376.

14.   Frances R. Curcio, ed., *Curriculum and evaluation standards for school mathematics: Addenda series grades 5 - 8* (Reston, VA: National Council of Teachers of Mathematics, 1991).

15.   Miriam A. Leiva, ed., *Curriculum and evaluation standards for school mathematics: Addenda series grades K - 6* (Reston, VA: National Council of Teachers of Mathematics, 1992).

16.   NCTM, *Professional standards for teaching mathematics* (Reston, VA: National Council of Teachers of Mathematics, 1991).

17.   Douglas B. McLeod et al., "Setting the Standards," in *Bold ventures: Case studies of U.S. innovations in mathematics education*, ed. Senta Raizen and Edwards D. Britton (Dordrecht: Kluwer Academic Publishers, 1996): 376.

18.   University of Illinois, *Undergraduate programs* (Urbana-Champaign, IL: University of Illinois, 1991).

19.   DESE, *Teacher certification* [online] [cited October 1 2000]). Available http://services.dese.state.mo.us/divurbteached/teachcert/certification 158.

20.   NCATE, "Program for initial preparation of K-4 teachers with an emphasis in mathematics, 5-8 mathematics teachers, 7-12 mathematics teachers," (Reston, VA: National Council of Teachers of Mathematics, October 1998): 415-420.

21.   Association for Childhood Education International, *ACEI position paper: Preparation of elementary teachers* [online][cited October 1 2000]). Available www.udel.edu/bateman/acei/prepel

22.   Douglas B. McLeod et al., "Setting the Standards," in *Bold ventures: Case studies of U.S. innovations in mathematics education*, ed. Senta Raizen and Edwards D. Britton (Dordrecht: Kluwer Academic Publishers, 1996): 376.

23.   Michael J. Bosse, "Reforming the NCTM Standards in light of historical perspective: Premature changes," *Journal of Mathematical Behavior* 17, no. 3 (1998): 317-327.

24.   Katharine Doyle Rasch, "The evolution of selected "modern" mathematics content in elementary school mathematics textbooks 1963-1982" (Doctor of Philosophy, St. Louis University, 1983).

25.   R. L. Wilder, "Development of modern mathematics," in *National Council of Teachers of Mathematics Yearbook* (Washington, D.C.: National Council of Teachers of Mathematics, 1969)

26.   Jeremy Kilpatrick, "Confronting reform," *American Mathematical Monthly* 104 (1997): 955-962.

27.   Sherman K. Stein, *Strength in numbers: Discovering the joy and power of mathematics in everyday life* (New York: John Wiley, 1996).

28.   Frances R. Curcio, ed., *Curriculum and evaluation standards for school mathematics: Addenda series grades 5 - 8* (Reston, VA: National Council of Teachers of Mathematics, 1991).

29.   Stephen S. Willoughby, *Mathematics education for a changing world* (Alexandria, Virginia: ASCD, 1990).

30.   Mathematically Correct, *Glossary of terms* [online] [cited May 12 2000]). Available http://mathematicallycorrect.com/glossary.htm

31.   NAEP (National Assessment of Educational Progress), *The nation's report card* [online] [cited October 1 2000]. Available http://nces.ed.gov/nationreportcard/site/home.asp

32.   NCTM, "Mathematics making a living, making a life," (Richmond, VA: National Council of Teachers of Mathematics).

33.   NCTM, *Assessment standards for school mathematics* (Reston, VA: National Council of Teachers of Mathematics, 1995).

34.   NACOME, *Overview and analysis of school mathematics, grades K-12*, ed. National Advisory Committee on Mathematical Education Conference Board of the Mathematical Sciences (Reston, VA: NCTM, 1975).

35.   Peter L. Glidden, "Teaching applications: will the pendulum of reform swing too far?," *The Mathematics Teacher* 89, no. 6 (1996): 450-451.

36.   Edwin E. Moise et al., *Five views of the "New Math"* (Washington, D.C.: Council for Basic Education, 1965). (Maryville University School of Education Archives).

Chapter VI: Conclusion

1.   Max A. Sobel and Evan M. Maletsky, *Teaching mathematics*, 3rded. (Boston: Allyn and Bacon, 1999).

2.   Jeremy Kilpatrick, *Five lessons from the new math era* [online] (Symposium of "Reflecting on Sputnik", Oct. 4, 1997 [cited July 22, 2000]). Available: www.nas.edu/sputnik/

3.   Theodore R. Sizer, "Reform movement or panacea," (Record Series 10/13/1, Box 3, University of Illinois Archives).

4.   Mathematically Correct, *Glossary of terms* [online] [cited May 12 2000]). Available http://mathematicallycorrect.com/glossary.htm

5.   Michael J. Bosse, "The NCTM Standards in light of the new math movement: A warning!," *Journal of Mathematical Behavior* 14, no. 2 (1995): 171-201.

6.   CEMREL, "Comprehensive school mathematics program: basic program plan," (St. Ann, MO: CEMREL, 1971).

7.   Decker Walker, *Fundamentals of curriculum* (Orlando: Harcourt Brace Jovanovich, 1990).

8.   TIMSS, *Third International Mathematics and Science Study* [online] [cited October 2 2000] Available http://nces.ed.gov/timss/, 2000.

9.   Douglas B. McLeod et al., "Setting the Standards," in *Bold ventures: Case studies of U.S. innovations in mathematics education*, ed. Senta Raizen and Edwards D. Britton (Dordrecht: Kluwer Academic Publishers, 1996): 376.

10.   NCTM, "The revolution in school mathematics," (Washington, D.C.: A report of Regional Orientation Conferences in Mathematics, 1961): 1-90.

11.   Michael J. Bosse, "Reforming the NCTM Standards in light of historical perspective: Premature changes," *Journal of Mathematical Behavior* 17, no. 3 (1998): 317-327.

12.   Glenda Lappan, *Lessons from the Sputnik era in mathematics education* [online] (Symposium of "Reflecting on Sputnik," Oct. 4, 1997 [cited July 22, 2000]). Available www.nas.edu/sputnik/

13.   NCTM, *Parent's corner* [online] [cited October 1, 2000]) Available www.nctm.org.

14.   Jeremy Kilpatrick, *Five lessons from the new math era* [online] (Symposium of "Reflecting on Sputnik", Oct. 4, 1997 [cited July 22, 2000]). Available: www.nas.edu/sputnik/

15.   Thomas Steven Dupre, "The University of Illinois Committee on School Mathematics and the "new mathematics" controversy" (Doctor of Philosophy, University of Illinois, 1986).

16.   Robert B. Davis, "A modern mathematics program as it pertains to the interrelationship of mathematical content, teaching methods and classroom atmosphere," (Webster Groves, MO

Index